Modeling and Optimization for Mobile Social Networks

Zhou Su · Qichao Xu · Kuan Zhang
Xuemin (Sherman) Shen

Modeling and Optimization for Mobile Social Networks

Zhou Su
School of Mechatronic Engineering and Automation
Shanghai University
Shanghai
China

Qichao Xu
School of Mechatronic Engineering and Automation
Shanghai University
Shanghai
China

Kuan Zhang
Department of Electrical and Computer Engineering
University of Waterloo
Waterloo, ON
Canada

Xuemin (Sherman) Shen
Department of Electrical and Computer Engineering
University of Waterloo
Waterloo, ON
Canada

ISBN 978-3-319-83859-5 ISBN 978-3-319-47922-4 (eBook)
DOI 10.1007/978-3-319-47922-4

Softcover reprint of the hardcover 1st edition 2016

Printed on acid-free paper

This Springer imprint is published by Springer Nature
The registered company is Springer International Publishing AG
The registered company address is: Gewerbestrasse 11, 6330 Cham, Switzerland

Preface

With the rapid development of mobile communication technologies and the ever-increasing population of users, mobile social network (MSN) has become an emerging paradigm of the next-generation network. The basic tenet of MSNs is that mobile users can form communities at any time and anywhere to exchange information with each other based on their interests, social tie, social activities, etc. Therefore, MSNs can facilitate a myriad of attractive applications related to social and communication services, including content delivery, community activities, and infotainment.

Due to the advance of social applications, it is necessary for MSNs to provide quality of experience (QoE), which poses new challenges of modeling and optimization in MSNs as follows: (1) there is a huge amount of information to be disseminated over MSNs. The characteristics of users during information dissemination should be studied including the social tie among users, interests in information, etc. An analytical model to optimize the information dissemination is needed; (2) with the new demands of social networks applications such as crowd sensing, users are encouraged to participate in MSNs to not only receive content, but also contribute the content. As some users may be selfish, an incentive model is necessary to encourage users to actively participate the MSNs; (3) in MSNs, the content-delivery path between the source and the destination is unstable or even unavailable sometimes. A store-carry-forward pattern is applied to deliver content to the destination in MSNs, where this delivery fashion needs the optimal relay services; (4) the resources such as the available bandwidth in MSNs are limited. Since the emerging social network applications may have different requirements of resources, the optimal bandwidth allocation becomes important to provide users with the satisfied mobile network services; and (5) various users take different types of mobile devices to access the MSNs through different wireless connections. A model based on the heterogeneous network architecture is needed to optimize the network architecture for MSNs.

In this monograph, we study the modeling and optimization for MSNs, since it is of great importance to address the above challenges and facilitate newly developed applications. In Chap. 1, an overview of MSNs is provided, by specifying the critical challenges, research issues, and emerging applications in MSNs. In Chap. 2 with the introduction of MSN-based social graph and information dissemination mechanism, the modeling and optimization for epidemic information dissemination in MSNs are discussed. In Chap. 3, by the analysis of the selfishness division of users, we present the modeling and optimization of selfishness-aware incentive to encourage users to participate cooperation in MSNs. In Chap. 4, based on the store-carry-forward fashion of content delivery, we present an optimal relay service for users. In Chap. 5, to deliver social multimedia through clouds, we study the modeling and optimization for cloud resource allocation in MSNs. In Chap. 6, by introducing small cells and macro-cell, the modeling and optimization for heterogeneous MSN architecture are investigated to deliver social contents in a content-centric mode. Finally, in Chap. 7, we present some open issues and future research directions in MSNs. In summary, this monograph validates the feasibilities of modeling and optimization for MSNs including epidemic information dissemination, selfishness-aware incentive, relay services, cloud resource allocation, and heterogeneous network architecture. It also evaluates the performance of the above studies. Therefore, this monograph provides valuable insights on the design and deployment of future MSNs.

We would like to thank Prof. Pinyi Ren at the Department of Communication and Information of Xi'an Jiaotong University, Xi'an, China; Prof. Song Guo at the School of Computer Science and Engineering of the University of Aizu, Aizu, Japan; Prof. Mianxiong Dong at the Department of Information and Electronic Engineering of Muroran Institute of Technology, Muroran, Japan; Prof. Yuan Wu at the School of Information Engineering of Zhejiang University of Technology, Zhejiang, China; Dr. Shan Zhang, at the Department of Electrical and Computer Engineering of University of Waterloo, Waterloo, Canada for their contributions in the presented research works. We would like to thank all of the members of BBCR group, University of Waterloo, Canada, for their valuable discussions, insights, and helpful comments. We would also like to thank the staff at Springer Science+Business Media: Ms. Susan Lagerstrom-Fife and Ms. Jennifer Malat, for their kind help throughout the publication and preparation processes.

Shanghai, China — Zhou Su
Shanghai, China — Qichao Xu
Waterloo, Canada — Kuan Zhang
Waterloo, Canada — Xuemin (Sherman) Shen

Contents

Acronyms

ACK	Acknowledgement Character
AP	Access Point
BS	Base Station
CCC	Credit Clearance Center
CCCDF	Cooperative Cache-based Content Dissemination Framework
CCNs	Content-Centric Networks
CS	Content Store
D2D	Device-to-device
FIB	Forwarding Information Base
GPS	Global Positioning System
GSM	Global System for Mobile Communications
H-MSNs	Healthcare-oriented Mobile Social Networks
ID	Identification
IoT	Internet of Things
IP	Internet Protocol
MCS	Mobile Crowd Sensing
M-MSNs	Multimedia-oriented Mobile Social Networks
MRR	Media Response Ratio
MSIoTs	Mobile Social Internet of Things
MSNs	Mobile Social Networks
ODEs	Ordinary Differential Equations
OSN	Online Social Network
PDAs	Personal Digital Assistants
PIT	Pending Interest Table
QoE	Quality of Experience
QoP	Quality of Protection
QoS	Quality of Service
RRA	Random Resource Allocation
SCC	Social-Centric Caching
SI	Susceptible Infectious

SNS	Social Network Service
TS	Time Stamp
TTL	Time-to-live
URA	Uniform Resource Allocation
XMPP	Extensible Messaging and Presence Protocol

Chapter 1
Introduction

Mobile social networks (MSNs) have become new network paradigms which combine mobile networks and social services based on the social relations among users. As both the population of users and the scale of networks are increasing, modeling and optimization for MSNs bring critical issues to satisfy the new demands of quality of experience (QoE) from users. In this chapter, we first provide an overview of MSNs. Then, we highlight the latest research directions related to the modeling and optimization for MSNs. Finally, the aim of the monograph is provided.

1.1 Overview of MSNs

1.1.1 Mobile Social Networks

Nowadays, users have been more disposed to obtain content by mobile devices such as mobile phones and tablets instead of the existing desktop computers. It has been reported that the number of mobile devices used by users has been more than the population in the world since 2014. As more and more people use mobile devices to access the Internet, various mobile social networks emerge to provide users with the satisfied network connections and social services. Some emerging social network applications, such as Facebook, Twitter in the North America and Wechat, SinaBlog in China, etc., have been designed and widely used in daily life.

In MSNs, different users may have different social activity based on his social characteristics and the relationship between other users. A group of users are linked up by one or multiple interdependent relationships. These interdependent relationships can be the physical contacts, common interests, financial trading activities, or organizational involvement [1]. Users with one or multiple interdependent relationships can form a community to contact with others through mobile devices [2].

Z. Su et al., *Modeling and Optimization for Mobile Social Networks*,
DOI 10.1007/978-3-319-47922-4_1

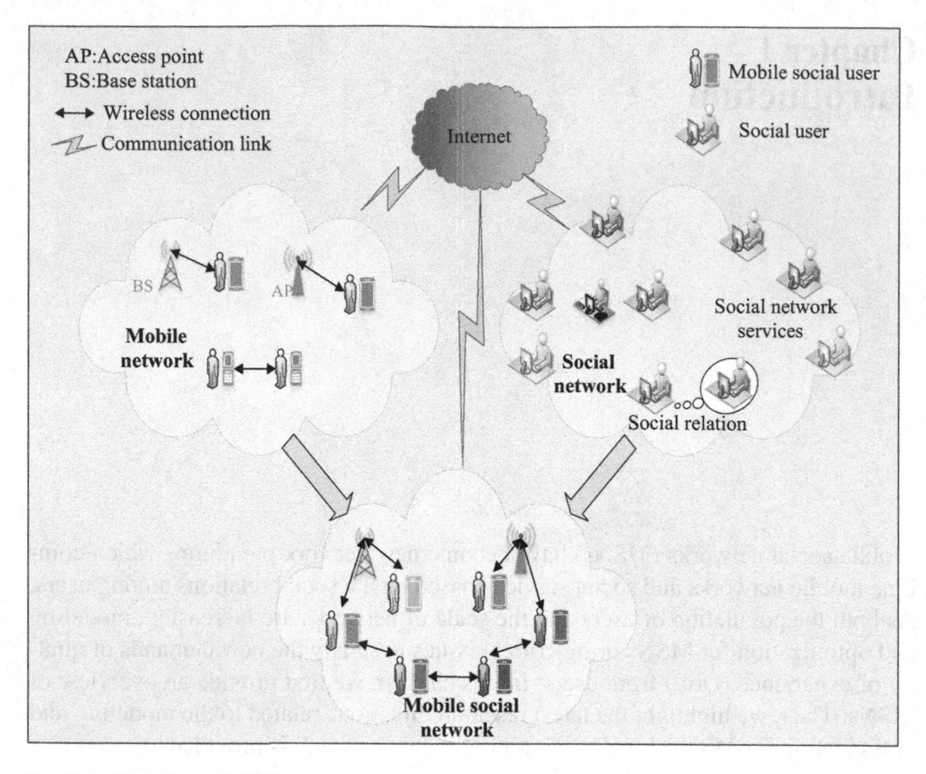

Fig. 1.1 Overview of MSNs

The MSNs provide users with the platform to carry out social activities instead of using the traditional computer. The mobility of social users makes the network services such as content exchange, sharing and transmission distinct from other conventional networks. The MSNs can be looked as an integrated paradigm of social networks and mobile communication networks, as shown in Fig. 1.1. During the delivery of service in MSNs, the QoE needs to be guaranteed with the consideration of optimization of parameters status such as the mobility, bandwidth and delay.

1.1.2 Architecture of MSNs

In the traditional communication way, the users obtain the data from the servers which store the data. However, MSNs require an increasing number of large-sized content delivery (e.g., audio or video contents), which may not be guaranteed with a satisfied QoE only through the traditional wireless communication networks. To this end, the short range wireless technologies (e.g. Wi-Fi and Bluetooth) can be leveraged and widely used in MSNs, as shown in Fig. 1.2. With these technologies,

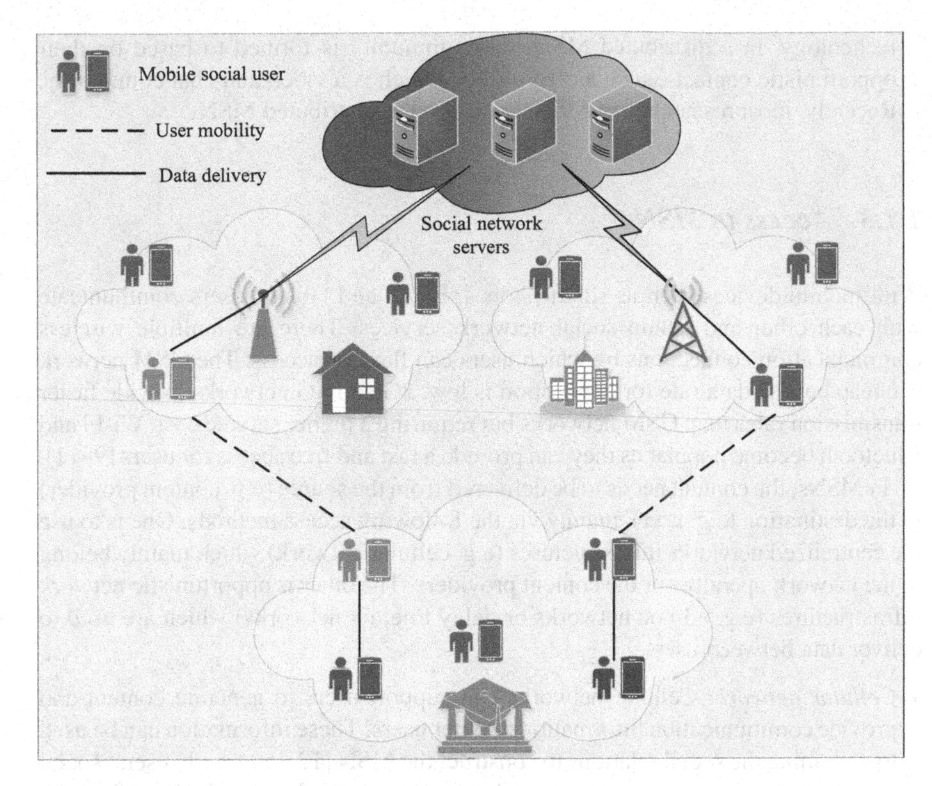

Fig. 1.2 User mobility and data delivery in MSNs

users can exchange data through mobile devices directly [3–7] when they encounter with each other. The structure of MSNs can be divided into two types, i.e., Web-based MSN and distributed MSN.

- *Web-based MSN*. Web-based MSN is a centralized structure which relies on network infrastructures and content servers. Social users use mobile devices through social network services (e.g., Facebook, Myspace and Twitter) or mobile portal (e.g., m.4info.com) to communicate with the Web-based applications. There have been many Web-based MSNs to provide social services for mobile users. For example, "Iphone Facebook App" is a Web-based MSN which allows users to communicate with the "Facebook". Similarly, "WhozThat" [2] uses the online information to obtain users' characteristics and interest to recommend music to the users. "Google Latitude" [8] is another Web-based MSNs. It can help users obtain the location information of others who have the will to share the location information.
- *Distributed MSN*. In a distributed MSN, users form a community to disseminate data between different social users without the central servers. The opportunistic contact between two users is defined as a link. When a user encounters another user, they can exchange or share the data directly through short-range wireless

technology. In a distributed MSN, the community is formed to based on their opportunistic contacts, such as "EyeVibe" which is a video and chat community. Recently, most researches of MSNs focus on the distributed MSN.

1.1.3 Access to MSNs

With mobile devices such as smart phones, PDAs and tablets, users communicate with each other and obtain social network services. There are multiple wireless communication connections by which users can flexibly access. The GSM network is cheap but the data rate for connection is low. 3G and 4G networks provide faster transmission rates than GSM networks but requiring a higher service cost. Wi-Fi and Bluetooth become popular as they can provide a fast and free access for users [9–11].

In MSNs, the content needs to be delivered from the source (e.g. content provider) to the destination (e.g. user), mainly via the following access methods. One is to use the centralized networks infrastructures (e.g. cellular network) which mainly belong to the network operators or the content providers. The other is opportunistic network infrastructures (e.g. ad-hoc networks or delay tolerant networks) which are used to deliver data between users.

- *Cellular network*: Cellular networks can support users to generate content and provide communication information to other users. These information can be used to determine the social relations to construct the MSN [12], and study users' social activities. Because the social relations between users are dynamic in the real world, the communication records are mapped into a time sequence with the correlation model. The model has been used to quantize the strength of the social relation between users to analyze the social relation among users.
- *Opportunistic network*: Due to the mobility of users, the communication in MSNs between the source user and destination user is dynamic without a fixed topology. For example, when users use Wi-Fi or Bluetooth to connect with each other, the opportunistic networks can be formed for MSNs. Because of the characteristics such as intermittent connectivity and long time disconnect, the opportunistic networks can support a store-carry-forward way to spread data in MSNs. The store-carry-forward way allows relay users to store the data when there is no chance to transmit it. Then the relay users transmit the data to the users who are close to the destination user by opportunistic contacts. When the relay user encounters another user, the relay user decides whether to forward the data to the encountering node or not by estimating the probability that this user will encounter the destination user [13]. Figure 1.3 shows the process that a source user transmits the data to the destination user in MSNs in opportunistic networks. Here, the source user A needs to transmit the data to user E which is the destination user. As the distance between user A and E is long to transmit the data from user A to user E directly, user A needs to find the relay users to transmit data to E. In this figure, user A selects user B and C as the relay users. Then user C transmits the data to user D who finally transmits the data to E.

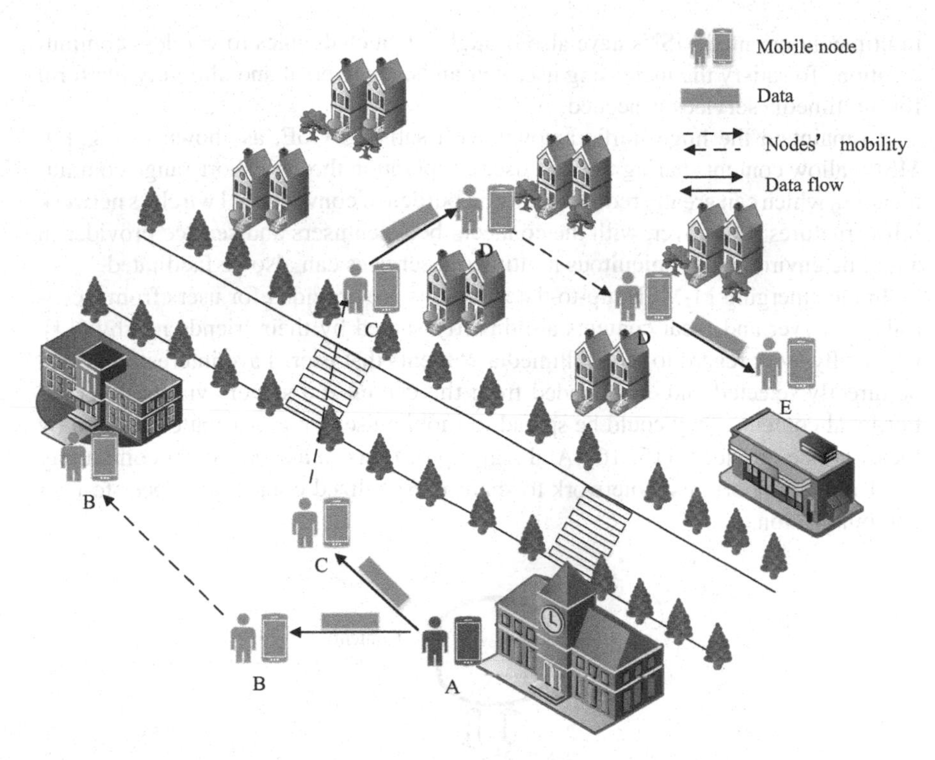

Fig. 1.3 Opportunistic network

1.2 Applications in MSNs

There have been many applications to provide various services to mobile social uses with MSNs. We discuss several applications in the following subsections, i.e., multimedia services, healthcare services, and security services.

1.2.1 *Multimedia Services*

Multimedia has become object-oriented and content-driven applications with user cooperation in recent years. According to the statistics in 2012, in every minute per day, there are 100,000 tweets posted; 48 h of videos in total updated on Youtube with 2,800,000 views, 685,000 pieces of contents shared on Facebook; and 2,000,000 queries searched on Google. The combination of MSNs and multimedia services has significantly changed users life styles, and fostered a value-added research direction of the multimedia-oriented mobile social network (M-MSN). However, the multimedia-oriented MSNs have also brought great challenges to wireless communication. To satisfy the increasing user demands, a universal and almighty platform for multimedia services is needed.

To maintain the huge traffic amount with satisfied QoE, as shown in Fig. 1.4, MSNs allow content sharing through user cooperation through short-range communication, which can greatly reduce the traffic burden of conventional wireless network infrastructures. Moreover, with the connects between users and service provider in a mobile environment, ubiquitous multimedia services can also be facilitated.

In the emerging M-MSN, up-to-date contents are provided for users from a centralized server and local contents are directly shared by their friends nearby [14]. Especially, with regard to the multimedia contents that users have interest, they can be directly selected and downloaded from the centralized servers via the Internet. For local contents, they could be spread to mobile users in a distributed fashion by local service providers [15, 16]. And a group of users in a same social community can form an opportunistic network to share personalized contents via peer-to-peer communication.

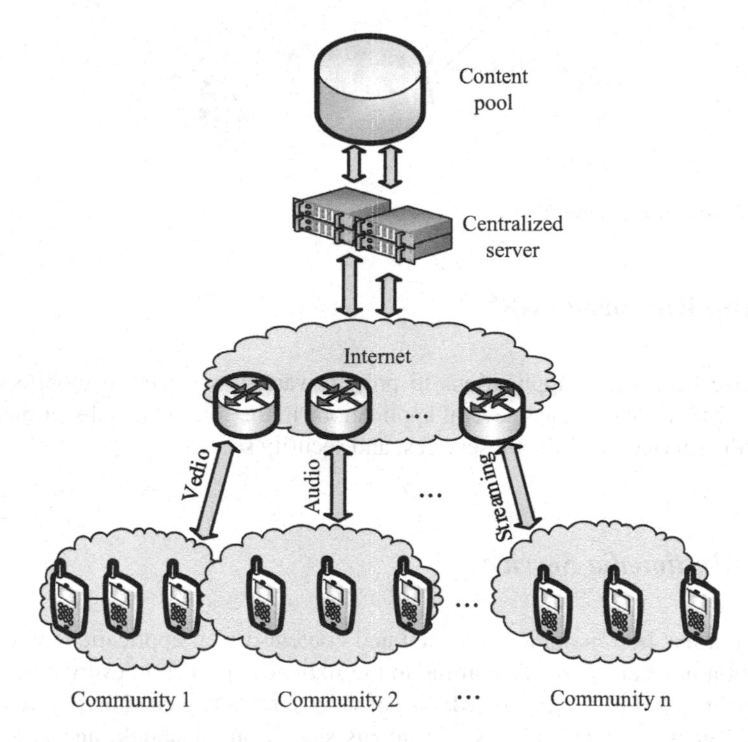

Fig. 1.4 Multimedia-oriented mobile social network

1.2.2 Heathcare Services

Healthcare, as one of the major economic and social issues around the world, is of great importance to an aging society. It requires tremendous labor resources and health expense. According to the 2014 national health report in the United States, the average expense for each person was $8895, while the annual national healthcare expenditure has increased to $3.8 trillion. The personal care, home health care, and nursing home care are about 18 % of tne the total expenditure [17, 18]. Furthermore, traditional hospital-centric healthcare always needs long waiting time with a low efficiency to identify serious diseases in the early stages or deal with chronic diseases. Thus, to develop the up-and-coming healthcare solutions, such as health monitoring as well as health data processing and sharing, will be very helpful to improve disease diagnosis and reduce the health expenditure [19].

Recently, wearable devices (e.g., smart rings, wristwatches, bracelets, and hair caps) have attracted wide attention and are widely applied to provide healthcare service, including calorie burn during fitness, physiology parameter monitoring [19], and heart rate recording. Healthcare-oriented mobile social networks (H-MSNs) has emerged, and consists of heterogeneous mobile networks (e.g., WiFi, cellular network and device-to-device [D2D] communications), the ubiquitous wearable devices, and powerful servers (e.g., cloud servers). As shown in Fig. 1.5, the H-MSNs can be

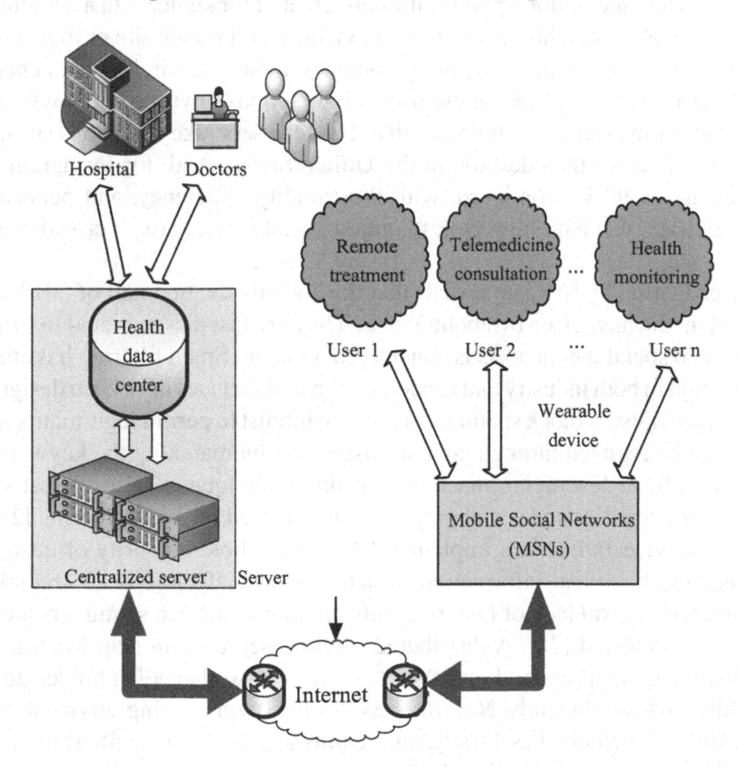

Fig. 1.5 Mobile healthcare network

used to collect the health information sensed by wearable devices, analyze the data of health diagnosis and monitoring, and enable the social interactions among users. For example, by wearing the dedicated wearable devices, seniors can measure their physiology information, such as body temperature, heart rate, and oxygen saturation. In the meantime, these health data can be remotely obtained by their family members and/or doctors with the use of desktops and smartphones via H-MSNs. For emergency case such as a heart problem or falling down, the patients physical situation can be automatically sent by the wearable devices to his/her family members and doctors. H-MSNs also allow users to share health condition and activity information sensed by wearable devices among friends [20].

1.2.3 Security Services

The MSNs, which can easily be constituted by smartphone users in a local area, has become a hopeful social networking platform to enable media sharing, social gaming, group chat, and interaction among users nearby [21]. Through WiFi, Bluetooth, and device-to-device communications, connections are established among users and an opportunistic network is formed for a long or temporary period (e.g., several hours). For instance, rich interaction opportunities are created for students in a campus area, residents in an urban neighborhood, tourists visiting or a scenic site or museum, and customers in a shopping mall. Without worrying about the wireless data charge as well as Internet access, MSNs allow users' interactions anytime and anywhere. A recent report from comScore indicates that Twitter users take over 86 % of time on mobile devices rather than desktop in the United States. And, for Instagram users, the percentage is 98 %. In a word, with the rapidity, efficiency, and pervasion of MSNs [22], users can have high QoE to obtain social services (e.g., crowdsourcing) [23, 24].

An investigation by Nexgate reveals that during only the first half of 2013, spams over social media have risen by about 355 %. They are fast disseminated in MSNs as every 1 of 200 social media posts is considered as spam. Spam filtering has attracted much attention in both industry and academic. Several services have been designed on the basis of blacklist to block spammers [25] or whitelist to permit legitimate senders. Filtering can be realized through content inspection by matching the keyword with the packets [26] or leveraging machine learning techniques [27] to detect spams. Relevant characteristics and social graph are also studied for spam filtering [28, 29].

Most of services which are implemented by a credible authority or centralized server need the historical information to detect spams. If there is no credible and centralized servers and lack of historical information in MSNs, spammers are more likely to be undetected [30]. A distributed filtering servcie can help MSN users to personalize the spam filters, and send these spam filters to others filter holders to allow them to filter spams efficiently. Nevertheless, for the spam filtering service in MSNs, there are still some challenges. Firstly, taking into account both the filtering accuracy and distribution costs, how to distribute filters is a challenge. Secondly, some filter creators' sensitive information, such as lifestyles, physical situation and preferences [26, 31], may be remained in the distributed filters. The privacy and security problems should be considered. In addition, some malicious attackers may forge the original filters to block some useful information, how to resist these malicious attackers is the third challenge.

1.3 Research Issues in MSNs

There have been many efforts to improve the performance of MSNs. We discuss several important research directions in the following subsections, i.e., community detection, content distribution, mobility, privacy and security.

1.3.1 Community Detection

Community detection is to discover communities within which users have the common interests or close social relationship. Some existing methods [32, 33] can be used to divide MSNs into some communities. However, as most of these methods are not designed based on users' social features, they are still inefficient to detect the unknown communities in MSNs. Furthermore, due to the mobility of users and the large scale of MSNs, the community detection in MSNs becomes more difficult. Therefore, there have been many studies for the community detection in MSNs.

Yang et al. [34] propose a semi-supervised community detection framework, in which the prior information is used to penalize the latent space dissimilarity of different nodes to detect communities in a social network. The proposed framework can improve the accuracy of community detection for networks with unclear structures. Lu et al. [35] focus on the community detection in weighted networks and propose a community detection method based on the strength of the relationship between users in the network. The authors define the intra-centrality and inter-centrality to characterize relationship between users in the network. Jia et al. [36] define the edge centrality to discriminate the edges between networks nodes. Based on an edge centrality, a novel algorithm called edge antitriangle centrality is proposed for community detection.

Whang et al. [37] propose an overlapping community detection method based on a seed expansion approach. The proposed method can not only produce cohesive clusters but also identify ground-truth communities by greedily expanding seeds with a community metric. Chen et al. [38] discuss two fine-tuned community detection algorithms to measure the community quality where the given network community structure can be split and merged. Su et al. [39] propose a fuzzy modularity maximization method for overlapping community detection, which is based on a modularity maximization algorithm. The globally optimal solution can be found by a tree-based structure.

1.3.2 Content Distribution

When users encounter with each other through opportunistic contacts, they may exchange content by using mobile devices directly, without network infrastructures. To distribute content from the source user to the destination, the appropriate relay nodes should be selected to deliver content, with considerations such as bandwidth utilization, interval time, interest in the content, etc. [40–46]. Besides, to further improve the QoE, the environment aware is also needed in MSNs to analyze and predict users' social features, based on users' environment information (e.g. location, conversation time etc.) [47–51].

To efficiently distribute content, Wu et al. [51] study the social features of users during the information propagation in MSNs. Two classes of selfishness are considered which are individual selfishness and social selfishness, respectively. Xu et al. [52] propose an incentive scheme to select the relay node in MSNs to overcome the problems caused by the users' selfishness. The problem of selection is formulated as a bargain game and the Nash equilibrium is obtained to incentive users to relay the content. Mei et al. [44] propose two forwarding protocols for MSNs which are Give2Get epidemic forwarding and Give2Get delegation forwarding, respectively. In the two protocols, the number of copies of data is limited to reduce the total overhead.

Bulut et al. [53] consider the friendship between users in MSNs and define a new metric to find the direct and indirect friends. Based on the friendship information in MSNs, the authors propose a friendship-based routing scheme for the content delivery. Xu et al. [54] study the epidemic information dissemination in MSNs and develop an analytical model to analyze dissemination process. In the model, the authors adopt preimmunity and immunity to represent the attitude about whether the user is interested in the information or not. Wang et al. [55] propose a cloud-based multicast scheme for MSNs, in which the information forwarding process is divided into two phases which are pre-cloud and inside-cloud.

1.3.3 Mobility

Different from the conventional social network where users use the desktop computers, users take mobile devices to access MSNs [56–58]. As users may change their locations during moving, the network topology and communication status become dynamic. Mobility of users plays a crucial role in the performance of social network applications such as content sharing, spreading and search. To design the efficient protocols and algorithms in MSNs, the mobility features should be studied.

Lin et al. [59] propose a random waypoint model for MSNs, in which each user selects a random destination and moves to the destination by a random speed. By considering the distance and speed, [59] can obtain an accurate derivation of distribution functions for the steady state. Kim et al. [60] develop a mobility model of users movements among Wi-Fi access points (APs). The authors collect the number of visitors on each AP and aggregate multiple days log into a single day to study the hourly arrival and departure rates on each AP. Lee et al. [61] use the semi-Markov process to develop a mathematical model to characterize both the steady-state and transient behaviors of users among APs. The steady-state behavior is to estimate the time that the user connects with an AP and the transient behavior is to predict the user's future locations.

Bettstetter et al. [62] study the node distribution by using the random waypoint mobility model and propose a general mobility model in which the pause time is arbitrarily chosen in the node's waypoint. The structure of distribution is shown by using three independent components which are the pause, static, and mobility component. Resta et al. [63] study the relation between nodes' mobility and the QoS. The authors propose a wireless QoS-aware mobility model which contains a user mobility model, a user traffic model, a wireless technology model and a QoS model. Balazinska [64] examine a trace with 1366 users and 117 APs over 4 weeks and focus on the study of population characteristics, load distribution among APs and node mobility.

1.3.4 Privacy

In spite of that the promising features of social networking services can be provided by MSNs, there exist new challenges [65]. Specially, as a typical multimedia service in an M-MSN, the personalized content query may expose a user's individual information to service providers. For example, a user's queries could be obtained and recorded by Google to analyze his/her concerns and preferences. Once users browse social networking sites and press the "Facebook Like" button, their individual information and preferences, such as identity and location, are exposed. Obviously, without a careful protection, users' privacy will be a critical issue in these applications. The privacy issues become more and more important [58, 66–70]. For example, the location privacy has become a serious concern for users when they use the location-based services. Besides, as autonomous mobile social applications have been easy to download nowadays, the user's personal preferences such as his friends nearby and the interested content, etc., may be revealed when using these autonomous mobile social applications. Therefore, there have been tremendous efforts to protect the privacy of users.

Zhu et al. [71] study to protect users' interest information to provide the secure friend discovery process in MSNs. They propose a privacy preserving and fairness-aware friend matching protocol. In the protocol, the blind vector transformation technique is used to protect users' interest by hiding the relationship of the original interest with the transformed vector. Li et al. [72] investigate the privacy protection in user profile matching and define two privacy levels. The authors propose two distributed privacy protection schemes for profile matching which are private set-intersection protocol and private cardinality of set-intersection protocol. Lu et al. [73] study the location privacy for users and propose a link-layer based location privacy protocol. As the protection is deployed at the link-layer, even the attacker gets the user's location information, the attacker cannot know how long the user has stayed in the current location.

Lu et al. [74] propose a privacy-preserving method for location privacy in MSNs which let users use pseudo-ID in the network. Because the pseudo-ID is frequently changed at different locations, the user's past, current and future locations can be protected. Wang et al. [75] propose a location-aware location privacy protection scheme. It allows users to define the dynamic and diverse privacy requirement according to different locations. Guo et al. [76] investigate the privacy problems in the content dissemination process by focusing the users' social information privacy and the dissemination content privacy. The authors propose a privacy-preserving social-assisted mobile content dissemination scheme. By using the verified identical attributes to establish users' potential social relationships, the proposed scheme can only provide the content to the appropriate users in a cryptographic way to protect the privacy of the user and the content.

1.3.5 Security

Security is one of the most important issues [21, 77] to be resolved in MSNs. There exist many kinds of attacks such as the Sybil attack, the flooding attack, etc. In addition, users may be cheated by malicious users who forge some contents in the system. For example, service evaluation applications enable users to share their experiences or reviews about services they are given. However, Sybil attackers may exist in this application and frequently change pseudonyms or identities to repeatedly broadcast the same or similar information [30, 78]. They may mislead the user's opinions and preferences [79]. Besides, the flooding attackers can exhaust the resources of users by transmitting mass data packets. As users have the limited resources such as the battery power, memory space, bandwidth and so on, how to propose security strategies against these attacks should be studied.

Zhang et al. [30] classify the Sybil attackers into four levels according to their attacking capabilities and propose a Sybil detection scheme to distinguish attackers in MSNs. Abbas et al. [80] present a lightweight Sybil attack detection method for MSNs, in which the entry and exit behaviors of all identities are studied to detect the Sybil identities. Yu et al. [81] propose a defense method against Sybil attacks called SybilGuard in which the social graph between the identities is used to detect the fake identities. Quercia et al. [82] identify Sybil attackers by allowing each mobile user to manage two small networks which are network of friends and network of foes, respectively. And the network information can be shared when two users have a contact.

Zhang et al. [83] study a flooding attack in MSNs where the malicious users broadcast mass data packets to other users and propose a generic defense with a threshold against the attacks. Yi et al. [84] discuss the influence of flooding attacks and analyze the network's performance under different degree of flooding attacks.

Kim et al. [85] propose a period based defense mechanism against the flooding attacks and each user in MSNs has a blacklist to prevent the attacks. Fallah et al. [86] consider the resources of both the normal users and the malicious users. They propose an optimum defense mechanism which can provide the maximum possible payoff for normal users based on the game theory.

1.4 Modeling and Optimization for MSNs

To address the aforementioned issues, massive efforts of modeling and optimization in MSNs are also needed. As one of the complex networks, MSNs consist of both centralized networks and distributed networks, resulting in complicated and dynamic network features and structures. The relationships in MSNs need to be modeled, from the perspectives of users, communities, and content providers. The abstractions of network parameters are considered as variables to show the social features of MSNs.

With the evolution of new mobile applications, the performance of existing MSNs need to be improved to satisfy the needs of mobile users. Optimizations with models can be used to maximize a certain output to improve the performance of MSNs. Especially, due to the dynamic network topology and limited resources, the modeling and optimization are needed for MSNs to regarding the efficiency of epidemic information dissemination, the cooperation degree of selfish-aware incentive mechanism, the selection of relay services, the profit of resource allocation, and the heterogeneous network architecture in MSNs.

1.5 Aim of the Monograph

Due to the emerging network applications and various demands from users, the modeling and optimization are needed to study. However, the existing research studies on the modeling and optimization in MSNs are limited, where the aforementioned unique features of MSNs should be taken into consideration. A deep study of modeling and optimization in MSNs will shed light not only to the design and implementation of optimal protocols for users, but also to some important economics issues, such as how to provide users with a satisfied service to reap profits from the view of content provider.

The aim of this monograph is to investigate how to model and optimize MSNs considering the unique features and different scenarios in MSNs. The modeling with optimization can be used to maximize a certain output in MSNs such as the efficiency of information dissemination, the cooperation degree of social users, the success of relay service, the profit of resource allocation, and the network architecture. Specifically, we try to answer the following research questions: (a) modeling and optimization for epidemic information dissemination in MSNs; (b) modeling and optimization for selfishness-aware incentive mechanism in MSNs; (c) optimal relay services in MSNs; (d) modeling and optimization for cloud resource allocation in MSNs; and (e) modeling and optimization for heterogeneous network architecture in MSNs. To answer these questions, in this monograph, we develop the corresponding models and optimizations considering the characteristics of MSNs. Based on the investigations on the above issues, it can help us elaborate the insights and implications to design and implement the future MSNs.

References

1. W. Gao, Q. Li, B. Zhao, G. Cao, Multicasting in delay tolerant networks: a social network perspective, in *Proceedings of the ACM MOBIHOC* (New York, 2009), pp. 299–308
2. V. Le, Z. Feng, D. Bourse, P. Zhang, A cell based dynamic spectrum management scheme with interference mitigation for cognitive networks. Wirel. Pers. Commun. **49**(2), 1594–1598 (2008)
3. H. Zhu, X. Lin, R. Lu, Y. Fan, Smart: a secure multilayer credit-based incentive scheme for delay-tolerant networks. IEEE Trans. Veh. Technol. **58**(8), 4628–4639 (2009)
4. K. Ota, M. Dong, J. Wang, S. Guo, Z. Cheng, M. Guo, Dynamic itinerary planning for mobile agents with a content-specific approach in wireless sensor networks, in *Proceedings of the IEEE VTC* (Ottawa, 2010), pp. 1–5
5. M. Dong, T. Kimata, K. Sugiura, K. Zettsu, Quality-of-experience (QoE) in emerging mobile social networks. IEICE Trans. Inf. Syst. **197**(10), 2606–2612 (2014)
6. W. Zhang, Y. Ye, H. Tan, Q. Dai, T. Li, Information diffusion model based on social network, in *Proceedings of the MCSAAISC* (Berlin, 2010), pp. 145–450
7. E. Bulut, B. Szymanski, Friendship based routing in delay tolerant mobile social networks, in *Proceedings of the IEEE GLOBECOM* (Miami, 2010), pp. 1–5
8. R. Pastor-Satorras, M. Rubi, A. Diaz-Guilera, Statistical mechanics of complex networks. Rev. Mod. Phys. **74**(1), 47–97 (2002)
9. N. Kayastha, D. Niyato, P. Wang, E. Hossain, Applications, architectures, and protocol design issues for mobile social networks: a survey. Proc. IEEE **99**(12), 2125–2129 (2011)
10. J. Sun, An incentive scheme based on heterogeneous belief values for crowd sensing in mobile social networks, in *Proceedings of the IEEE GLOBECOM* (Atlanta, 2013), pp. 1717–1722
11. Q. Xu, Z. Su, B. Han, D. Fang, Z. Xu, Analytical model for epidemic information dissemination in mobile social networks with a novel selfishness division, in *Proceedings of the LSMS and ICSEE* (Shanghai, 2014), pp. 469–475
12. S. Ali, J. Qadir, A. Baig, Routing protocols in delay tolerant networks—a survey, in *Proceedings of the ICET* (Islamabad, 2010), pp. 70–75
13. A. Karam, N. Mohamed, Middleware for mobile social networks: a survey, in *Proceedings of the HICSS* (Maui, 2012), pp. 1482–1490
14. P. Ha, P. Tsigas, J. Anshus, F. Sname, SocioNet: a social-based multimedia access system for unstructured P2P networks. IEEE Trans. Parallel Distrib. Syst. **21**(7), 1027–1041 (2010)
15. G. Cardone, A. Corradi, L. Foschini, R. Montanari, Socio-technical awareness to support recommendation and efficient delivery of ims-enabled mobile services. IEEE Commun. Mag. **50**(6), 82–90 (2012)
16. I. Roussaki, N. Kalatzis, N. Liampotis, P. Kosmides, M. Anagnostou, K. Doolin, E. Jennings, Y. Bouloudis, S. Xynogalas, Context-awareness in wireless and mobile computing revisited to embrace social networking. IEEE Commun. Mag. **50**(50), 74–81 (2012)
17. K. Zhang, X. Shen, *Security and Privacy for Mobile Healthcare Networks* (Springer, Berlin, 2015)
18. http://www.forbes.com/

19. K. Zhang, K. Yang, X. Liang, Z. Su, X. Shen, H. Luo, Security and privacy for mobile healthcare networks-from quality-of-protection perspective. IEEE Wirel. Commun. **22**(4), 104–112 (2015)
20. A. Toninelli, R. Montanari, A. Corradi, Enabling secure service discovery in mobile healthcare enterprise networks. IEEE Wirel. Commun. **16**(3), 24–32 (2009)
21. K. Zhang, X. Liang, X. Shen, R. Lu, Exploiting multimedia services in mobile social networks from security and privacy perspectives. IEEE Commun. Mag. **52**(3), 58–65 (2014)
22. K. Wei, M. Dong, K. Ota, K. Xu, Camf: Context-aware message forwarding in mobile social networks. IEEE Trans. Parallel Distrib. Syst. **26**(8), 2178–2187 (2015)
23. A. Azaria, A. Richardson, S. Kraus, V. Subrahmanian, Behavioral analysis of insider threat: a survey and bootstrapped prediction in imbalanced data. IEEE Trans. Comput. Soc. Syst. **1**(2), 135–155 (2014)
24. J. Ren, Y. Zhang, K. Zhang, X. Shen, SACRM: social aware crowdsourcing with reputation management in mobile sensing. Comput. Commun. **65**, 55–65 (2015)
25. F. Soldo, A. Le, A. Markopoulou, Blacklisting recommendation system: using spatio-temporal patterns to predict future attacks. IEEE J. Sel. Areas Commun. **29**(7), 1423–1437 (2011)
26. K. Zhang, X. Liang, R. Lu, X. Shen, PIF: a personalized fine-grained spam filtering scheme with privacy preservation in mobile social networks. IEEE Trans. Comput. Soc. Syst. **2**(3), 41–52 (2015)
27. B. Agrawal, N. Kumar, and M. Molle, Controlling spam emails at the routers, in *Proceedings of the IEEE ICC* (Seoul, 2005), pp. 1588–1592
28. Z. Li, H. Shen, SOAP: a social network aided personalized and effective spam filter to clean your e-mail box, in *Proceedings of the IEEE INFOCOM* (Shanghai, 2011), pp. 1835–1843
29. M. Sirivianos, K. Kim, X. Yang, SocialFilter: introducing social trust to collaborative spam mitigation, in *Proceedings of the IEEE INFOCOM* (Shanghai, 2011), pp. 2300–2308
30. K. Zhang, X. Liang, R. Lu, K. Yang, X. Shen, Exploiting mobile social behaviors for sybil detection, in *Proceedings of the IEEE INFOCOM* (Hong Kong, 2015), pp. 271–279
31. M. Li, H. Zhu, Z. Gao, S. Chen, L. Yu, S. Hu, K. Ren, All your location are belong to us: breaking mobile social networks for automated user location tracking, in *Proceedings of the ACM MOBIHOC* (Philadelphia, 2014), pp. 43–52
32. X. Sheng, J. Tang, X. Xiao, G. Xue, Leveraging GPS-less sensing scheduling for green mobile crowd sensing. IEEE Internet Things J. **1**(4), 328–336 (2014)
33. M. Newman, Detecting community structure in networks. Eur. Phys. J. B Condens. Matter Complex Syst. **38**(2), 321–330 (2004)
34. L. Yang, X. Cao, D. Jin, X. Wang, D. Meng, A unified semi-supervised community detection framework using latent space graph regularization. IEEE Trans. Cybern. **45**(11), 2585–2598 (2015)
35. Z. Lu, X. Sun, Y. Wen, G. Cao, Algorithms and applications for community detection in weighted networks. IEEE Trans. Parallel Distrib. Syst. **26**(11), 2916–2926 (2015)
36. S. Jia, L. Gao, Y. Gao, H. Wang, Anti-triangle centrality-based community detection in complex networks. IET Syst. Biol. **8**(3), 116–25 (2014)
37. J. Whang, D. Gleich, I. Dhillon, Overlapping community detection using neighborhood-inflated seed expansion. IEEE Trans. Knowl. Data Eng. **28**(5), 1272–1284 (2016)
38. M. Chen, K. Kuzmin, B. Szymanski, Community detection via maximization of modularity and its variants. IEEE Trans. Comput. Soc. Syst. **1**(1), 46–65 (2014)
39. J. Su, T. Havens, Quadratic program-based modularity maximization for fuzzy community detection in social networks. IEEE Trans. Fuzzy Syst. **23**(5), 1356–1371 (2015)
40. L. Danon, A. Dłazguilera, J. Duch, A. Arenas, Comparing community structure identification. J. Stat. Mech. Theory Exp. (2005)
41. M. Talasila, R. Curtmola, C. Borcea, Improving location reliability in crowd sensed data with minimal efforts, in *Proceedings of the WMNC* (Dubai, 2013), pp. 1–8
42. Y. Wen, J. Shi, Q. Zhang, X. Tian, Quality-driven auction-based incentive mechanism for mobile crowd sensing. IEEE Trans. Veh. Technol. **64**(9), 4203–4214 (2015)

43. L. Pournajaf, A. Garcia-Ulloa, L. Xiong, V. Sunderam, Participant privacy in mobile crowd sensing task management: a survey of methods and challenges, in *Proceedings of the ACM SIGMOD* (New York, 2014)
44. H. Sun, C. Wu, Epidemic forwarding in mobile social networks, in *Proceedings of the IEEE ICC* (Ottawa, 2012), pp. 1421–1425
45. Y. Wu, S. Deng, H. Huang, Information propagation through opportunistic communication in mobile social networks. Mob. Netw. Appl. **17**(6), 773–781 (2012)
46. A. Mei, J. Stefa, Give2Get: forwarding in social mobile wireless networks of selfish individuals. IEEE Trans. Dependable Secur. Comput. **9**(4), 568–581 (2012)
47. M. Karaliopoulos, Assessing the vulnerability of DTN data relaying schemes to node selfishness. IEEE Commun. Lett. **13**(12), 923–925 (2010)
48. Y. Li, G. Su, D. Wu, D. Jin, L. Su, L. Zeng, The impact of node selfishness on multicasting in delay tolerant networks. IEEE Trans. Veh. Technol. **60**(5), 2224–2238 (2011)
49. H. Zhu, L. Fu, G. Xue, Y. Zhu, M. Li, L. Ni, Recognizing exponential inter-contact time in vanets, in *Proceedings of IEEE INFOCOM* (New Jersey, 2010), pp. 1–5
50. T. Karagiannis, J. Boudec, M. Vojnovic, Power law and exponential decay of intercontact times between mobile devices. IEEE Trans. Mob. Comput. **9**(10), 1377–1390 (2010)
51. Y. Wu, S. Deng, H. Huang, Evaluating the impact of selfish behaviors on epidemic forwarding in mobile social networks. J. Stat. Mech. Theory Exp. **2013**(2), P02018 (2013)
52. Q. Xu, Z. Su, S. Guo, "A game theoretical incentive scheme for relay selection services in mobile social networks. IEEE Trans. Veh. Technol. **PP**(99), 1 (2015)
53. E. Bulu, B. Szymanski, Exploiting friendship relations for efficient routing in mobile social networks. IEEE Trans. Parallel Distrib. Syst. **23**(23), 2254–2265 (2012)
54. Q. Xu, Z. Su, K. Zhang, P. Ren, X. Shen, Epidemic information dissemination in mobile social networks with opportunistic links. IEEE Trans. Emerg. Top. Comput. **3**(3), 399–409 (2015)
55. Y. Wang, J. Wu, W. Yang, Cloud-based multicasting with feedback in mobile social networks. IEEE Trans. Wirel. Commun. **12**(12), 6043–6053 (2013)
56. X. Fan, V. Li, K. Xu, Fairness analysis of routing in opportunistic mobile networks. IEEE Trans. Veh. Technol. **63**(3), 1282–1295 (2014)
57. X. Zhang, Z. Zhang, J. Xing, R. Yu, Exact outage analysis in cognitive two-way relay networks with opportunistic relay selection under primary user's interference. IEEE Trans. Veh. Technol. **64**(6), 1–1 (2014)
58. K. Fall, A delay-tolerant network architecture for challenged internets, in *Proceedings of the ACM SIGCOMM* (New York, 2003), pp. 27–34
59. G. Lin, G. Noubir, R. Rajmohan, Mobility models for ad hoc network simulation, in *Proceedings of IEEE INFOCOM* (Hong Kong, 2004), p. 463
60. M. Kim, D. Kotz, Modeling users' mobility among wifi access points, in *Proceedings of the WTMM* (Berkeley, 2005), pp. 19–24
61. J. Lee, C. Hou, Modeling steady-state and transient behaviors of user mobility: formulation, analysis, and application, in *Proceedings of the ACM MOBIHOC* (New York, 2006), pp. 85–96
62. C. Bettstetter, G. Resta, P. Santi, The node distribution of the random waypoint mobility model for wireless ad hoc networks. IEEE Trans. Mob. Comput. **2**(3), 257–269 (2003)
63. G. Resta, P. Santi, Wiqosm: an integrated qos-aware mobility and user behavior model for wireless data networks. IEEE Trans. Mob. Comput. **7**(2), 187–198 (2008)
64. M. Balazinska, P. Castro, Characterizing mobility and network usage in a corporate wireless local-area network, in *Proceedings of the ICMSAS* (New York, 2003), pp. 303–316
65. X. Liang, K. Zhang, X. Shen, X. Lin, Security and privacy in mobile social networks: challenges and solutions. IEEE Wirel. Commun. **21**(21), 33–41 (2014)
66. Y. Wang, M. Chuah, Y. Chen, Incentive based data sharing in delay tolerant mobile networks. IEEE Trans. Wirel. Commun. **13**(13), 370–381 (2014)
67. T. Chen, L. Zhu, F. Wu, S. Zhong, Stimulating cooperation in vehicular ad hoc networks: a coalitional game theoretic approach. IEEE Trans. Veh. Technol. **60**(2), 566–579 (2011)
68. F. Wu, T. Chen, S. Zhong, C. Qiao, G. Chen, A game-theoretic approach to stimulate cooperation for probabilistic routing in opportunistic networks. IEEE Trans. Wirel. Commun. **12**(4), 1573–1583 (2013)

69. L. Wei, Z. Cao, H. Zhu, Mobigame: a user-centric reputation based incentive protocol for delay/disruption tolerant networks, in *Proceedings of the IEEE GLOBECOM* (Houston, 2011), pp. 1–5
70. M. Mahmoud, X. Shen, PIS: a practical incentive system for multihop wireless networks. IEEE Trans. Veh. Technol. **59**(8), 4012–4025 (2010)
71. H. Zhu, S. Du, M. Li, Z. Gao, Fairness-aware and privacy-preserving friend matching protocol in mobile social networks. IEEE Trans. Emerg. Top. Comput. **1**(1), 192–200 (2013)
72. M. Li, S. Yu, N. Cao, W. Lou, Privacy-preserving distributed profile matching in proximity-based mobile social networks. IEEE Trans. Wirel. Commun. **12**(5), 2024–2033 (2013)
73. R. Lu, X. Lin, H. Zhu, P. Ho, X. Shen, A novel anonymous mutual authentication protocol with provable link-layer location privacy. IEEE Trans. Veh. Technol. **58**(3), 1454–1466 (2009)
74. R. Lu, X. Lin, Z. Shi, J. Shao, PLAM: A privacy-preserving framework for local-area mobile social networks, in *Proceedings of the IEEE INFOCOM* (Toronto, 2014), pp. 763–771
75. Y. Wang, D. Xu, F. Li, Providing location-aware location privacy protection for mobile location-based services. Tsinghua Sci. Technol. **21**(3), 243–259 (2016)
76. L. Guo, C. Zhang, H. Yue, Y. Fang, A privacy-preserving social-assisted mobile content dissemination scheme in DTNS. IEEE Trans. Mob. Comput. **13**(12), 2301–2309 (2013)
77. Q. Lian, Z. Zhang, M. Yang, B. Zhao, An empirical study of collusion behavior in the maze p2p file-sharing system, in *Proceedings of the ICDCS* (Toronto, 2007), pp. 56–56
78. K. Yang, K. Zhang, J. Ren, X. Shen, Security and privacy in mobile crowdsourcing networks: challenges and opportunities. IEEE Commun. **53**(8), 75–81 (2015)
79. K. Zhang, X. Liang, R. Lu, X. Shen, Sybil attacks and their defenses in the internet of things. IEEE Internet Things J. **1**(5), 372–383 (2014)
80. S. Abbas, M. Merabti, D. Llewellyn-Jones, K. Kifayat, Lightweight sybil attack detection in manets. IEEE Syst. J. **7**(2), 236–248 (2013)
81. H. Yu, M. Kaminsky, P. Gibbons, A. Flaxman, Sybilguard: defending against sybil attacks via social networks. IEEE/ACM Trans. Netw. **16**(3), 576–589 (2008)
82. D. Quercia, S. Hailes, Sybil attacks against mobile users: friends and foes to the rescue, in *Proceedings of the IEEE INFOCOM* (San Diego, 2010), pp. 1–5
83. Y. Ping, H. Yafei, Z. Yiping, Z. Shiyong, D. Zhoulin, Flooding attack and defence in ad hoc networks. J. Syst. Eng. Electron. **17**(2), 410–416 (2006)
84. P. Yi, F. Zou, V. Zou, Z. Wang, Performance analysis of mobile ad hoc networks under flooding attacks. J. Syst. Eng. Electron. **22**(11), 334–339 (2011)
85. H. Kim, R.B. Chitti, J. Song, Novel defense mechanism against data flooding attacks in wireless ad hoc networks. IEEE Trans. Consum. Electron. **56**(2), 579–582 (2010)
86. M. Fallah, A puzzle-based defense strategy against flooding attacks using game theory. IEEE Trans. Dependable Secur. Comput. **7**(1), 5–9 (2010)

Chapter 2
Modeling of Epidemic Information Dissemination for MSNs

In this chapter, the modeling of epidemic information dissemination in MSNs is discussed. We introduce two new elements, i.e., pre-immunity and immunity. Then, we study the features of information dissemination for users. Based on the process of the epidemic information dissemination, a novel dissemination mechanism is proposed with four dissemination rules. An analytical model is also developed through ordinary differential equations to mimic epidemic information dissemination.

2.1 Information Dissemination in MSNs

The MSNs [1–4] enable users to exchange and share information via opportunistic peer-to-peer links with short range wireless communication techniques, such as Wi-Fi and Bluetooth [5–8]. With the opportunistic links in MSNs, users adopt store-carry-and-forward mode to disseminate information [9, 10]. When a user moves to the communication coverage area of others, the information can be successfully forwarded; when the user is out of the communication range of others, this user stores and carries the information and waits for the next connection with others.

As users receive information by employing opportunistic contacts (as shown in Fig. 2.1), the user's mobility has a significant impact on information dissemination [11, 12]. However, since the opportunistic peer-to-peer links cannot be easily observed, the procedures of information dissemination are unpredictable. Therefore, a model to analyze the epidemic information dynamics of MSNs is necessary to efficiently capture the realistic features of information dissemination in MSNs.

Although many research studies [13–15] have investigated the analytical model for wireless networks, most of them assume that the transmission path between two nodes is stable so that they cannot be directly applied to MSNs. Moreover, the social ties among users have impacts on information dissemination in MSNs. For example, a

Z. Su et al., *Modeling and Optimization for Mobile Social Networks*,
DOI 10.1007/978-3-319-47922-4_2

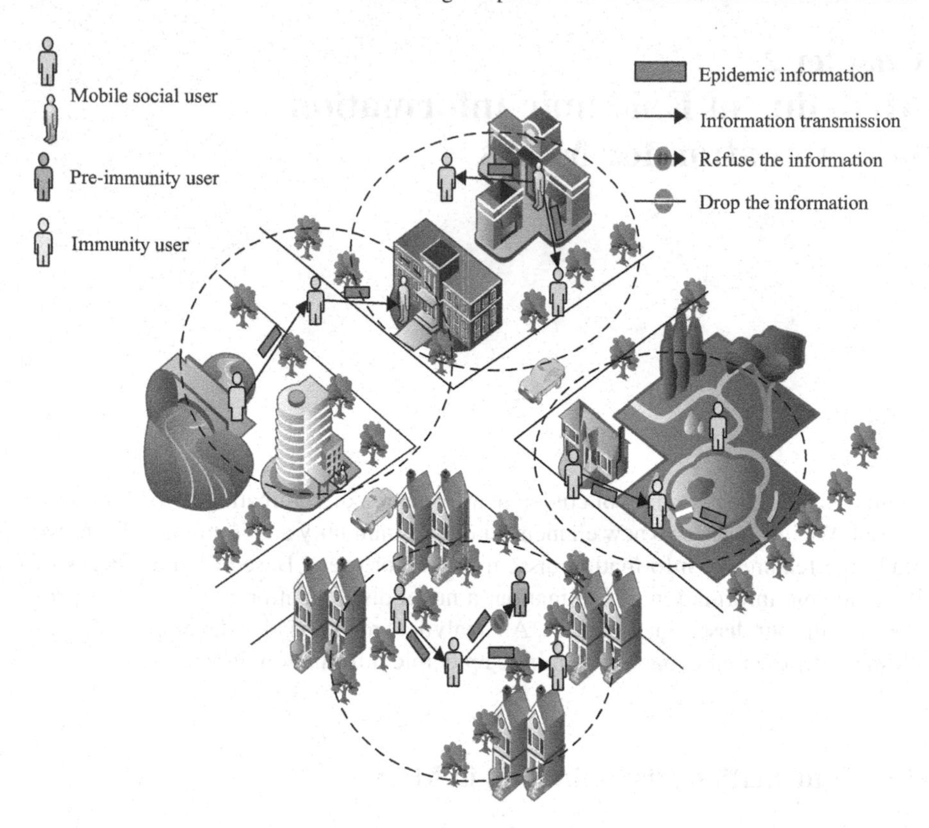

Fig. 2.1 Epidemic information dissemination among users

user may be willing to help forward information from his friends or the users with high social ties instead of a stranger. It is preferred to select social friends to store-carry-and-forward the information other than the users with high mobility. Therefore, how to balance the social tie and mobility becomes a challenging issue for information dissemination in MSNs. In addition, the user's interests may also vary during the information dissemination in MSNs. For example, in a local commercial street, a user has interests in food or Italian restaurant during the lunch time, while he may be interested in clothes and shoes after lunch. As the user changes his interests or social preferences, he would be willing to forward the current interested information rather than the one with old interests. Therefore, it is very important to develop an analytical model to investigate information dissemination in MSNs with the consideration of social impacts.

2.2 Related Work

2.2.1 Epidemic Information Dissemination Model

In the conventional epidemic information dissemination model, users have susceptible, infected and recovered status. Their states can be denoted by *S*, *I* and *R*, respectively. Here, *S* denotes the susceptible individuals. *I* denotes the infected individuals who have already obtained the information. *R* means the ones who are recovered and will not spread information.

The SI model [16] consists of *S* and *I* users. The infected users, who spread the information to the susceptible one with a given probability, are the source of the epidemic information spreading. Because *S* users get infectiousness after being infected, only *I* users exist at the end of information spreading. The SIS model [17] also has *S* and *I* users. In this model, *I* users can be recovered and then become *S* users, who may be infected when they contact *I* users.

Compared with the above models, the SIR [17] model can have *R* users. If *S* users are recovered, they will obtain the immunity and change to be *R* users who do not spread information. In the SIR model, the fraction of infected users grows gradually firstly. Then, the density of *I* users keeps decreased until 0 due to the low number of *S* users as time passes. The main difference between SIR model and other models is that there is no infected user in SIR model after information spreading.

All of the above three models focus on capturing the relations among users. They assume that any of the infected user has the possibility to infect other users. However, in real life an *S* user only infects his friends with the information. Therefore, in MSNs, the user who is to forward the information can only infect his neighbors when they are still not infected.

SCIR model [18] improves the SIR model by further considering the status that a user receives information and has no plan to forward it immediately. This status can also be changed to be infected or refractory. In the SCIR model, the nodes mean the user and the edges denote the relation between them. Zhang et al. [19] shows an information spreading model based on epidemics dynamics, which is an improved SIR model. Based on the topology of nodes and the effect on information spreading, the probability that a susceptible user becomes infectious after getting the information is studied, and the probability that an infected user becomes a recovered user after contacting recovered users is also analyzed. In addition, infected nodes can stop forwarding information with a certain speed.

With the development of mobile networks, more and more information is delivered over the users. Information spreading dynamically varies under the mobile environment, where the mobility of users [20], available bandwidth and communication frequency may affect the performance of spreading. Therefore, new studies related to the information spreading model in MSNs should be given.

2.2.2 General Information Dissemination Model

In traditional social networks, Zou et al. [14] present an Internet e-mail worm simulation model and find that the topology of network has the impact on the spread of e-mail worms in social networks. Wang et al. [15] study a rumor dissemination mode in complex social networks, which is similar to epidemic dissemination among humans. Nekovee et al. [21] introduce a rumor dissemination model in complex social networks and discuss the threshold behavior and dynamics of the model in different networks such as random graphs, scale-free networks, etc. Zhang et al. [19] propose an information dissemination model in online social networks, which considered the degrees of nodes and epidemiology. Zhang et al. [22] present an improved susceptible infectious (SI) model to mimic information spreading in online social networks, by taking the relationship of topology into consideration.

There are also several existing works about information dissemination in mobile networks. Dang and Wu [23] present a cluster-based routing protocol for delay-tolerant mobile networks, where mobile nodes with similar mobility pattern are gathered into a cluster to share limited resources. Taipov et al. [24] propose an efficient data sharing scheme called discover-predict-deliver in delay tolerant smartphone networks, which utilizes a mobility learning algorithm and a hidden Markov model to provide mobility information of individuals. Sammou et al. [25] exploit history contact and frequency of visiting different zones of networks to improve routing protocol. Ma and Jamalipour [26] propose a cooperative cache-based content dissemination framework (CCCDF) to provide caching and cooperative requesting strategies.

Recently, there have been an increasing number of studies on MSNs [27, 28]. Most of these studies focus on content distribution. Nazir et al. [5] present a content delivery scheme to perform relay selection by considering both the mobility patterns and encounter time of users. Using this scheme, content delivery can be optimized with lower end-to-end delay in time critical applications. Bulut and Szymanski [29] present a friendship-based routing scheme to make the forwarding decision with a novel metric which can help each node to define its friendship. By introducing a novel concept called destination cloud, Wang et al. [30] propose a scheme to provide multicasting services with feedback control mechanism in MSNs. Costa et al. [31] present a routing framework called SocialCast for publish-subscribe MSNs, which exploits predictions based on social interaction metrics to identify the best information carrier.

Several existing works investigate modeling information dissemination in MSNs. Sun and Wu [32] present a social aware epidemic forwarding model in MSNs, which studied the end-to-end delivery delay and considered both the limited and unlimited message validity in models. Wu et al. [33] model epidemic-like information dissemination in MSNs with selfish nodes, where the number of hops is limited. AduGyamfi et al. [34] model passive worms spreading in MSNs and carry out the analysis to find an effective antidote to fight against passive worms. Wu et al. [35] present a basic model and an extended model to evaluate the performance of information, where information can be shared between any pair of nodes whether they are friends or not.

Although the existing literature has studied several interesting aspects of MSNs, in most of the aforementioned and other related works, the model reflecting the dynamical social features of mobile nodes in the information has not been thoroughly studied. In this chapter, we develop an analytical model to analyze epidemic information in MSNs, which considers the dynamical characteristics of social features of mobile nodes.

2.3 System Model

In this section, we present the system model consisting of three parts: MSN-based social graph, opportunistic links and information dissemination mechanism.

2.3.1 MSN-Based Social Graph

We consider N mobile nodes (or users) take smartphones to bi-directionally communicate with others and self-organize an MSN, where there are some physical constraints [36] including battery and channel fading. Hence, it is impractical for each node to spread all information to all nodes. The mobile nodes in MSNs select a part of nodes as their friends and have social ties with them to spread information. An undirected social graph $G(V, E)$ is used to represent social ties among the mobile nodes, where V denotes the set of nodes, and E denotes the set of edges among nodes. Nodes in $G(V, E)$ can be the users or devices in MSNs. An edge exists between two nodes if they have a social tie, e.g., relatives, friends, and etc. For simplicity, two nodes which have social ties are considered to be friends in this chapter. The number of the nodes in MSNs is $|V| = N$.

To study the impact of social ties, we need to know the distribution of the friendship. Let $P(k)$ denote the probability that a node has degree k, where the degree of a node in the network refers to the number of its friends. The degree distribution $P(k)$ can be calculated by the fraction of nodes in the network with degree k. Thus, if there are N_k nodes having the degree k, we have $P(k) = N_k/N$.

Existing studies have shown that a great number of social networks have scale-free structures [37]. In other words, $P(k)$ conforms to a power-law distribution [32]:

$$P(k) \sim k^{(-\gamma)}, \gamma \in (2, 3), \tag{2.1}$$

where γ is called skewness of the degree distribution, and can be adjusted according to the scale of the network. Furthermore, the expectation of the degree distribution can be denoted by $\langle k \rangle = \sum_m^{N-1} kP(k)$, where m is the smallest degree of nodes in the network.

2.3.2 *Opportunistic Links*

Nodes in the MSN communicate with each other in a store-carry-and-forward mode with the short range communication techniques. Only when two nodes come within the transmission range, there is an opportunistic link between them to exchange information. Therefore, the mobility patterns of nodes have a significantly influence on the information dissemination in MSNs. Recently, some complicated models have been employed to predict the mobility patterns of nodes, such as Edge-Markov model [38]. However, due to the complexity, it is still difficult to use the above algorithms in practice. Thus, some simple mobility models are still required, such as [39, 40], although the performance cannot be completely guaranteed. Most of these models indicate that the inter-contact time between two nodes conforms to an exponential distribution [41]. Karagianis et al. [42] also shows that an exponential decay of inter-contact time between mobile devices is reasonable. Therefore, we assume that the inter-contact time between two nodes follows an exponential distribution with parameter λ in this chapter. The probability that two nodes encounter with each other within $[t, t + \Delta t]$ becomes

$$P(T \leq \Delta t) = \int_0^{\Delta t} \lambda e^{-\lambda t} dt = 1 - e^{-\lambda \Delta t}. \tag{2.2}$$

2.3.3 *Information Transmission Mechanism*

To introduce the information transmission mechanism clearly, nodes in the network can have three states: ignorant, spreading and recovered. A node is ignorant if it is interested in the information but has not yet received it. Nodes that have possessed a copy of information and are willing to disseminate the information to others can be seen as spreading nodes. If a node is not interested in the information and not willing to disseminate it either, this node can be looked upon as a recovered node.

Combined with three states of mobile nodes mentioned above, the detailed information dissemination process is introduced. Usually, users may be willing to help their friends rather than anyone upon contact, which is a practical concern in the real world but ignored in most of previous studies. In MSNs, we consider that there is only one node having the information initially, which is the spreading node. All of other nodes are interested in the information at first, and willing to receive the information. However, most of the nodes cannot always maintain the same interest. Some ignorant nodes may lose the interest later, and refuse to receive it. In other words, an ignorant node can directly become a recovered node, which is called pre-immunity. An information delivery occurs from one node to another node only when they are friends and encounter with each other. In the real world, a spreading node cannot successfully deliver the information to its friends all the time due to the constraints of QoS during data transmission. A spreading node successfully forwards the information to one of its ignorant friends with a certain probability. In addition, a spreading node may stop dissemination when it encounters a recovered friend-node, which is called immunity. Moreover, spreading nodes may stop dissemination without any contacts due to their disinclination to deliver the information.

According to the information dissemination process, we summarize the following four rules.

1. When a spreading node meets an ignorant friend-node, the ignorant node receives the information from the spreading node and becomes a spreading node with probability β, which is called spreading parameter.
2. An ignorant node may lose interests and directly become a recovered node. We assume that this change is based on an exponential distribution with parameter μ, where this assumption is also used in [35].
3. When a spreading node meets a recovered friend-node, the spreader becomes a recovered node with probability δ, which is called immune parameter.
4. Due to the decrease of interest in information, the spreading node may cease delivering information and become a recovered node spontaneously without meeting other nodes. The spreading node becomes recovered node based on an exponential distribution with the parameter ν.

Parameters μ and ν are the self-immune parameter of ignorant nodes and spreading nodes, respectively. According to these four dissemination rules, the state transition diagram of a node can be shown by Fig. 2.2.

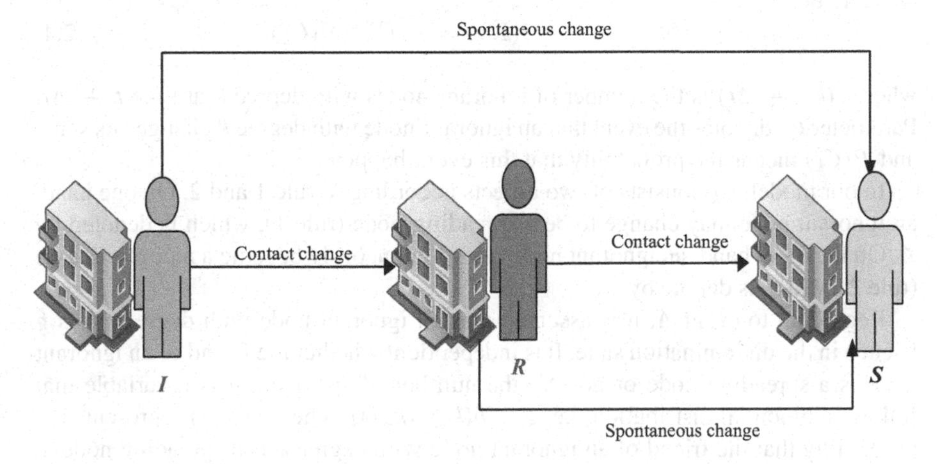

Fig. 2.2 State transition diagram of a node. Symbols I, S, and R represent the ignorant, spreading and recovered state, respectively

2.4 Proposed Analytical Scheme for Information Dissemination

In this section, we develop the analytical model through ordinary differential equations (ODEs) to mimic epidemic information dissemination in an MSN with N nodes, which can move around all the time in a region. The information can be delivered from one node to another only when they encounter and are socially connected with each other. The objective of this chapter is to study the success rate of the information transmission in MSNs, which is the total number of nodes receiving the information over time.

Let $I(k,t)$, $S(k,t)$, and $R(k,t)$ denote the number of ignorant nodes, spreading nodes, and recovered nodes with degree k at time t, respectively. The number of nodes with degree k at time t is represented by $N(k,t)$. In addition, $i(k,t)$, $s(k,t)$ and $r(k,t)$ are defined as the fraction of $I(k,t)$, $S(k,t)$ and $R(k,t)$ in $N(k,t)$, respectively. So we have

$$\begin{cases} i(k,t) = I(k,t)/N(k,t) \\ s(k,t) = S(k,t)/N(k,t) \\ r(k,t) = R(k,t)/N(k,t) \end{cases}, \tag{2.3}$$

where $i(k,t) + s(k,t) + r(k,t) = 1$.

According to the dissemination mechanism, given by the time interval $[t, t+\Delta t]$, we can obtain

$$i(k,t+\Delta t) - i(k,t) = -i(k,t)P(C_k). \tag{2.4}$$

where $i(k,t+\Delta t)$ is the number of ignorant nodes with degree k at time $t+\Delta t$. Parameter C_k denotes the event that an ignorant node with degree k changes its state, and $P(C_k)$ means the probability that this event happens.

In our model, C_k consists of two aspects according to rule 1 and 2. On one hand, an ignorant node may change to be a spreading node (rule 1), which is denoted by A. On the other hand, an ignorant node can spontaneously become a recovered node (rule 2), which is denote by B.

Regarding to event A, it is assumed that an ignorant node with degree k has g friends in the dissemination state. It is independent whether the friend of an ignorant node is a spreading node or not. So the number of spreaders g is a variable that follows a binomial distribution, i.e. $g \sim b(k, w(k,t))$, where $w(k,t)$ represents the probability that the friend of an ignorant node with degree k is a spreading node at time t.

To derive $w(k,t)$, two steps are considered as follows. First, the probability that an ignorant node with degree k has a link with a node with degree k' should be calculated. Second, we obtain the probability that this node is a spreader connected by the ignorant node with degree k. According to [21],

$$w(k, t) = \sum_{k'} P(k'|k)P(s_{k'}|i_k), \tag{2.5}$$

where $P(k'|k)$ denotes the probability that a node with degree k' is the neighbor of a node with degree k. Here, $P(s_{k'}|i_k)$ is defined as the probability that a node with degree k' is a spreading node, under the condition that it connects to an ignorant with degree k. For simplicity, $w(k, t)$ can be approximately written as [14]:

$$w(k, t) \approx \sum_{k'} P(k'|k)s(k', t), \tag{2.6}$$

where $s(k', t)$ denotes the fraction of degree-k' spreading nodes in degree-k' nodes at time t.

Let C_1 denote the event that an ignorant node encounters one of its friends who are in the dissemination state within $[t, t + \Delta t]$. According to (2.2), the probability that this event happens is

$$P(C_1) = 1 - \int_0^{\Delta t} \lambda e^{-\lambda x} dx = 1 - e^{-\lambda \Delta t}. \tag{2.7}$$

The event that an ignorant node receives the information from its dissemination neighbor under the condition that these two nodes have encountered each other is denoted by S_1. The probability of this event is given by

$$P(S_1|C_1) = \beta. \tag{2.8}$$

Combining (2.7) and (2.8), we can obtain the probability that an ignorant node does not receive the information (referred to event NM_1) in the time interval $[t, t+\Delta t]$ as

$$P(NM_1) = 1 - (1 - e^{-\lambda \Delta t})\beta. \tag{2.9}$$

Therefore, the probability $P(A)$ that an ignorant node with degree k becomes a spreader within $[t, t + \Delta t]$ is given by

$$\begin{aligned} P(A) &= 1 - P(\overline{A}) \\ &= 1 - \sum_{g=0}^{k} \binom{k}{g} w(k, t)^g \left(1 - w(k, t)\right)^{k-g} \left(P(NM_1)\right)^g \\ &= 1 - \sum_{g=0}^{k} \binom{k}{g} \left(w(k, t)P(NM_1)\right)^g \left(1 - w(k, t)\right)^{k-g} \\ &= 1 - \left(1 - w(k, t) + w(k, t)P(NM_1)\right)^k \\ &= 1 - \left(1 - w(k, t) + w(k, t)\left(1 - (1 - e^{-\lambda \Delta t})\beta\right)\right)^k. \end{aligned} \tag{2.10}$$

Next, we discuss event B that an ignorant node changes its state to the recovered state spontaneously within $[t, t + \Delta t]$. Let $P(\overline{B})$ denote the probability that an ignorant node remains ignorant. According to information dissemination rule (2), we have

$$P(\overline{B}) = 1 - \int_0^{\Delta t} \mu e^{-\mu t} dt = e^{-\mu \Delta t}. \tag{2.11}$$

Therefore, by combining (2.10) with (2.11), we have

$$\begin{aligned} P(C_k) &= 1 - P(\overline{A})P(\overline{B}) \\ &= 1 - (1 - w(k,t) + w(k,t)P(NM_1))^k \, e^{-\mu \Delta t} \\ &= 1 - \left(1 - w(k,t) + w(k,t)\left(1 - (1 - e^{-\lambda \Delta t})\beta\right)\right)^k e^{-\mu \Delta t}. \end{aligned} \tag{2.12}$$

where the term of $P(\overline{A})P(\overline{B})$ denotes the probability that the ignorant node does not change its state within $[t, t + \Delta t]$. Thus, the probability that the ignorant node changes its state to be spreading or recovered node is $1 - P(\overline{A})P(\overline{B})$.

Based on (2.4), the derivative of $i(k,t)$ becomes

$$\begin{aligned} \frac{\partial i(k,t)}{\partial t} &= \lim_{\Delta t \to 0} \frac{i(k, t + \Delta t) - i(k,t)}{\Delta t} \\ &= -i(k,t) \lim_{\Delta t \to 0} \frac{P(C_k)}{\Delta t}. \end{aligned} \tag{2.13}$$

According to (2.12), in the limit $\Delta t \to 0$, we obtain

$$\lim_{\Delta t \to 0} P(C_k) = \lim_{\Delta t \to 0} 1 - (1 - \theta)^k e^{-\mu \Delta t} = 0, \tag{2.14}$$

where $\theta = w(k,t) - w(k,t)(1 - (1 - e^{-\lambda \Delta t})\beta)$ and $\lim_{\Delta t \to 0} \theta = 0$.

Therefore, the right side of (2.13) conforms to L'Hospital's rule [43]. Thus we can have

$$\begin{aligned} \lim_{\Delta t \to 0} \frac{P(C_k)}{\Delta t} &= \lim_{\Delta t \to 0} \frac{1 - (1 - \theta)^k e^{-\mu \Delta t}}{\Delta t} \\ &= \lim_{\Delta t \to 0} \lambda \beta k w(k,t)(1 - \theta)^{k-1} e^{-\lambda \Delta t} e^{-\mu \Delta t} + \mu e^{-\mu \Delta t}(1 - \theta)^k \\ &= \lambda \beta k w(k,t) + \mu. \end{aligned} \tag{2.15}$$

Combining (2.6), (2.13), and (2.15), we have

$$\begin{aligned} \frac{\partial i(k,t)}{\partial t} &= -\lambda \beta k i(k,t) w(k,t) - \mu i(k,t) \\ &= -\lambda \beta k i(k,t) \sum_{k'} P(k'|k) s(k',t) - \mu i(k,t), \end{aligned} \tag{2.16}$$

For the spreading nodes, we have

$$s(k, t + \Delta t) - s(k, t) = i(k, t)P(A) - s(k, t)P(D_k). \tag{2.17}$$

where D_k denotes the event that a spreading node with degree k becomes a recovered node within $[t, t+\Delta t]$. The probability that this event happens is $P(D_k)$. The variation of the number of spreading nodes is caused by two reasons. One is that the ignorant nodes meet the dissemination friend-nodes and may become the spreading nodes. The other is that the spreading nodes meet the recovered friend-nodes and become the recovered nodes or their states may be spontaneously changed to be recovered. There is an increase in $s(k, t)$ due to the former aspect, while the latter decreases $s(k, t)$.

Similar to C_k, event D_k comprises two aspects according to the information dissemination rule (2) and rule (3). At first, a spreading node may change its status to be recovered (rule 3), which is denoted by event E. In addition, a spreading node can spontaneously become a recovered node (rule 4) denoted by event F.

About event E, we assume that a spreading node with degree k has l friends in the recovered state. Similarly, l is a binomial random variable $l \sim b(k, q(k, t))$, where $q(k, t)$ is the probability that a friend of the spreading node with degree k is a recovered node at time t. Then, we can obtain

$$q(k, t) \approx \sum_{k'} P(k'|k)r(k', t). \tag{2.18}$$

The event that a spreading node meets a friend in the recovered state within $[t, t + \Delta t]$ is denoted by C_2 with a probability as

$$P(C_2) = \int_0^{\Delta t} \lambda e^{-\lambda x} dx = 1 - e^{-\lambda \Delta t}. \tag{2.19}$$

Let $P(S_2|C_2)$ denote the probability that the spreading node becomes a recovered one, under the situation that this node has encountered a recovered friend-node in $[t, t + \Delta t]$. Thus, we have

$$P(S_2|C_2) = 1 - \delta. \tag{2.20}$$

According to (2.19) and (2.20), the probability $P(NM_2)$ that a spreader does not convert to a recovered node in $[t, t + \Delta t]$ becomes

$$P(NM_2) = 1 - (1 - e^{-\lambda \Delta t})\delta. \tag{2.21}$$

The probability $P(E)$ that a spreading node with degree k becomes a recovered node within $[t, t + \Delta t]$ is

$$
\begin{aligned}
P(E) &= 1 - P(\overline{E}) \\
&= 1 - \sum_{l=0}^{k} \binom{k}{l} q(k,t)^l \left(1 - q(k,t)\right)^{k-l} \left(P(NM_2)\right)^l \\
&= 1 - \left(1 - q(k,t) + q(k,t)P(NM_2)\right)^k \\
&= 1 - \left(1 - q(k,t) + q(k,t)\left(1 - (1 - e^{-\lambda \Delta t})\delta\right)\right)^k .
\end{aligned} \tag{2.22}
$$

Then, we discuss event F that a spreading node becomes a node in the recovered state spontaneously in $[t, t+\Delta t]$. According to dissemination rule (4), by neglecting event E, the probability $P(F)$ that the spreading node stays in the dissemination state in $[t, t+\Delta t]$ is

$$
P(\overline{F}) = 1 - \int_0^{\Delta t} \nu e^{-\nu t} dt = e^{-\nu \Delta t}. \tag{2.23}
$$

Combining (2.22) and (2.23), we have

$$
\begin{aligned}
P(D_k) &= 1 - P(\overline{E})P(\overline{F}) \\
&= 1 - \left(1 - w(k,t) + w(k,t)P(NM_2)\right)^k e^{-\nu \Delta t} \\
&= 1 - \left(1 - w(k,t) + w(k,t)\left(1 - (1 - e^{-\lambda \Delta t})\delta\right)\right)^k e^{-\nu \Delta t} .
\end{aligned} \tag{2.24}
$$

Based on (2.11), (2.17), and (2.24), by setting $\Delta t \to 0$, we have

$$
\begin{aligned}
\frac{\partial s(k,t)}{\partial t} &= i(k,t) \lim_{\Delta t \to 0} \frac{P(A)}{\Delta t} - s(k,t) \lim_{\Delta t \to 0} \frac{P(D_k)}{\Delta t} \\
&= \lambda \beta k i(k,t) \sum_{k'} P(k'|k) s(k',t) - \lambda \delta k s(k,t) \sum_{k'} P(k'|k) r(k',t) - \nu s(k,t).
\end{aligned} \tag{2.25}
$$

For the recovered nodes, we have

$$
r(k, t+\Delta t) - r(k,t) = i(k,t)P(B) + s(k,t)P(D_k). \tag{2.26}
$$

By combining (2.11) and (2.24), it can be updated as

$$
\frac{\partial r(k,t)}{\partial t} = \lambda \delta k s(k,t) \sum_{k'} P(k'|k) r(k',t) + \nu s(k,t) + \mu i(k,t). \tag{2.27}
$$

To simplify the problem, by ignoring the correlation of degrees among nodes, the probability that an edge points to a spreading node is independent of the degree of the node from which the edge is emanating. From [44], we have

$$P(k'|k) = \frac{k'P(k')}{<k>}. \tag{2.28}$$

The ODEs of the densities of ignorant, dissemination and recovered nodes become

$$\frac{\partial i(k,t)}{\partial t} = -\lambda\beta k i(k,t)\sum_{k'}\frac{k'P(k')}{<k>}s(k',t) - \mu i(k,t), \tag{2.29}$$

$$\frac{\partial s(k,t)}{\partial t} = \lambda\beta k i(k,t)\sum_{k'}\frac{k'P(k')}{<k>}s(k',t) - \lambda\delta k s(k,t)\sum_{k'}\frac{k'P(k')}{<k>}r(k',t) - \nu s(k,t), \tag{2.30}$$

$$\frac{\partial r(k,t)}{\partial t} = \lambda\delta k s(k,t)\sum_{k'}\frac{k'P(k')}{<k>}r(k',t) + \nu s(k,t) + \mu i(k,t). \tag{2.31}$$

As $\frac{\partial i(k,t)}{\partial t} + \frac{\partial s(k,t)}{\partial t} + \frac{\partial r(k,t)}{\partial t} = 0$, the quantities satisfy the normalization condition where $i(k,t) + s(k,t) + r(k,t) = 1$.

We can obtain the number of nodes in the ignorant state at time t, which is denoted by $I(t)$.

$$I(t) = \sum_{k=m}^{N-1} P(k)i(k,t)N. \tag{2.32}$$

Define $S(t)$ and $R(t)$ as the number of nodes in the dissemination state and recovered state, respectively. We have

$$S(t) = \sum_{k=m}^{N-1} P(k)s(k,t)N. \tag{2.33}$$

$$R(t) = \sum_{k=m}^{N-1} P(k)r(k,t)N. \tag{2.34}$$

According to the above analysis, we can obtain following observations: (1) From (2.29), it can be known that the differential of $i(k,t)$ is negative. So the number of ignorant nodes keeps decreasing gradually. (2) In the earlier stage of the dissemination, there are a few spreading nodes and recovered nodes in the network. Thus, the right side of (2.30) is positive at this moment. However, with the increasing number of spreading nodes and recovered nodes, the differential of $s(k,t)$ becomes negative. Therefore, the number of spreading nodes firstly increases and then gradually decreases. (3) Since the right side of (2.31) is positive, the number of recovered nodes keeps increasing from 0.

Here, in our analytical model developed through the ordinary differential equations, the information is transmitted among nodes based on the distributed MSNs.

By introducing the aforementioned four rules, the process of epidemic information dissemination can be concisely presented, where the computational complexity can be also reduced.

2.5 Performance Evaluation

2.5.1 Simulation Setup

In this section, we conduct trace-driven simulation to validate the proposed analytic model, time evolutions of nodes, and performance of information dissemination. The trace data are from the logs of MSN system developed by Xi'an Jiaotong University, based on the XMPP Protocol. The MSN system is used for the students in the university. And the data for our simulation was taken during the period from Aug. 26, 2013 to Sept. 2, 2013. Based on the communication between two friends, a social graph can be generated by the above trace. We use the largest connected sub-graph of this social graph to conduct our experiments. The basic parameters of this graph are $N = 802$, $E = 1222$, where the average degree is 3.05. The largest degree and smallest degree in the social graph are 72 and 1, respectively, which indicates a large degree fluctuation. Since the existing empirical studies show that the average inter-contact time between two users is around 5 h [40], we determine that the parameter of the exponential distribution of inter-contact is $\lambda = 0.002$ with the unit time of the system as 0.01 h.

2.5.2 Model Validation

To validate the analytical model of epidemic information dissemination, we develop a simulation platform of the epidemic information dissemination in Matlab. Moreover, to compare with other conventional models, the number of infected nodes $F(t)$ is used as the metric, which is the number of nodes receiving information at time t. $F(t)$ can be calculated as follows in our model.

$$F(t) = S(t) + R(t) = \sum_{k=m}^{N-1} +P(k)(s(k) + r(k))N, \; for \, \mu = 0. \qquad (2.35)$$

At the beginning of the simulation, there is only one spreading node picked randomly, while all other nodes are ignorant. To obtain $F(t)$, we set $\mu = 0$, which denotes that all recovered nodes and spreading nodes have received information at time t. The time of information dissemination process T varies from 0 to 5000. For other parameters, we set $\beta = 0.7$, $\delta = 0.3$ [19], and $\nu = 0.0001$ [35]. We compare our model with the conventional models including epidemic forwarding model [32], information dissemination model in online social network [22], and the basic model

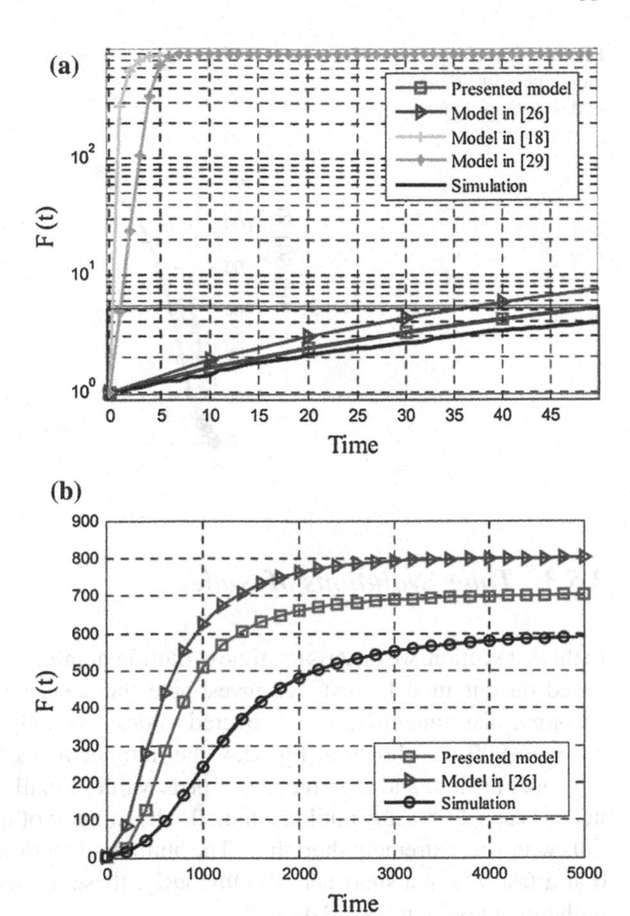

Fig. 2.3 Comparison of models and simulation. **a** Comparison in 45 time slots. **b** Comparison 5000 time slots

in [35]. Through 100 times of simulations, the results in Fig. 2.3 show that the difference between our presented model and the simulation result is the smallest. In addition, it is observed that the other three models have deviations, compared with our model. The reason is that the model in [32] ignores spreading parameter and makes information certainly to be delivered between two friend-nodes when they meet. Therefore, it is faster than our model to deliver information. The model in [22], where the friend-nodes are connected all the time, causes information to be forwarded to the entire network during an extremely short period. The basic model in [35] considers that a node can forward information to any node when they meet. In other words, there is no social relation in the model. Nevertheless, our model, taking into account the social nature and users' behaviors, is consistent with the simulation results. Accordingly, our model can be used to evaluate the effects of relevant parameters on information dissemination.

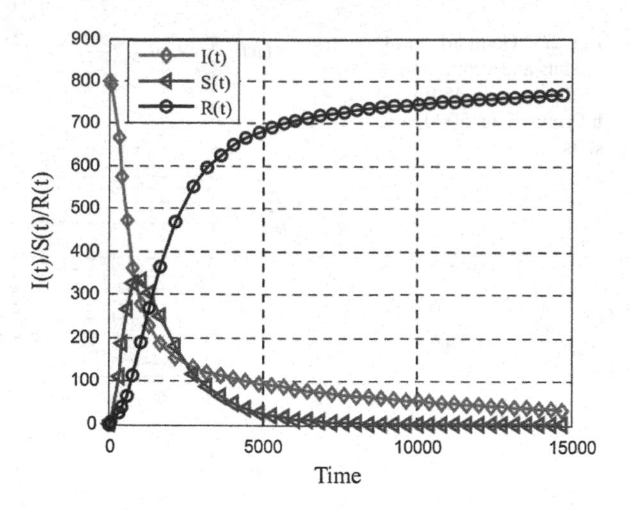

Fig. 2.4 Time evolutions of the number of ignorant nodes, spreading nodes and recovered nodes

2.5.3 *Time Evolutions of Nodes*

In this subsection, we discuss the time evolution of information dissemination process based on our model. First, we investigate the time evolutions of the number of ignorant, dissemination, and recovered nodes. We only change $\mu = 0.0001$ and $T = 15{,}000$. As shown in Fig. 2.4, the number of spreading nodes $S(t)$ sharply increases at first, and then reaches a peak value. Finally, the number of spreading nodes keeps decreasing until reaching 0. The number of ignorant nodes $I(t)$ sharply falls within an extremely short time. The number of recovered nodes $R(t)$ grows from 0 at a fast rate in a short period. Obviously, these numerical results are consistent with our analysis in Sect. 2.4.

We study the pre-immunity using the presented model. In Fig. 2.4, the value of $S(t)$ becomes around 0 when dissemination time $t = 10{,}000$, which indicates that there is no spreading node in the network. In other words, the information dissemination process is stopped at this moment. At the same time, there are still a few ignorant nodes in the network, which would lose the interest and become recovered nodes gradually. Obviously, the process of transition is considerably slow since the self-immune parameter μ is very small. For example, in Fig. 2.4, the value of $I(t)$ is reduced about 0.2 % in each unit of time when the value of $S(t)$ is almost 0. In order to further exhibit pre-immunity, we set $\mu = 0$ to study time evolutions of the number of three types of nodes without pre-immunity in Fig. 2.5. We can observe that three curves become stable when there is no spreading node in the network. The reason is that ignorant nodes maintain interests all the time if $\mu = 0$, and thus ignorant nodes cannot directly become recovered nodes.

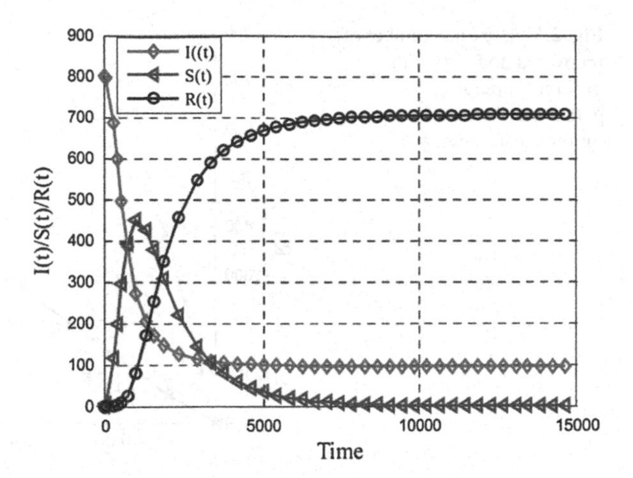

Fig. 2.5 Time evolutions of the number of ignorant nodes, spreading nodes, and recovered nodes, where $\mu = 0$

2.5.4 *Performance of Information Dissemination*

In this subsection, we evaluate the performance of information dissemination with relevant factors of the proposed model.

2.5.4.1 Effects of Spreading Parameter and Immune Parameter

We evaluate the effects of spreading parameter and immune parameter on the information dissemination. Here, $R(t)$ is used as the performance metric. In order to study the impact of spreading parameter and immune parameter, we ignore the self-immune parameter such that all recovered nodes can have the information under this situation. Here, the spreading parameter β increases from 0 to 1, and four values of δ are randomly chosen for comparison, which are 0.1, 0.3, 0.7, and 1. Other settings are the same as those in Fig. 2.5. Figure 2.6 shows the final number of the recovered nodes R versus different values of β and δ. From Fig. 2.6, it is observed that the information can be still transmitted if β is very small. For example, when $\beta = 0.1$ and $\delta = 0.3$, the number of the recovered node R is close to 200. Another phenomenon is that the final R increases with the increase of the value of β, while the increasing speed decreases gradually. Moreover, the larger immune parameter has smaller value of the final R. A large spreading parameter β can promote information dissemination, while immune parameter δ has a negative effect on information dissemination. Moreover, when δ is smaller, the network is more robust to β. For example, when $\delta = 0.1$, the value of final R remains stable when $\beta > 0.6$.

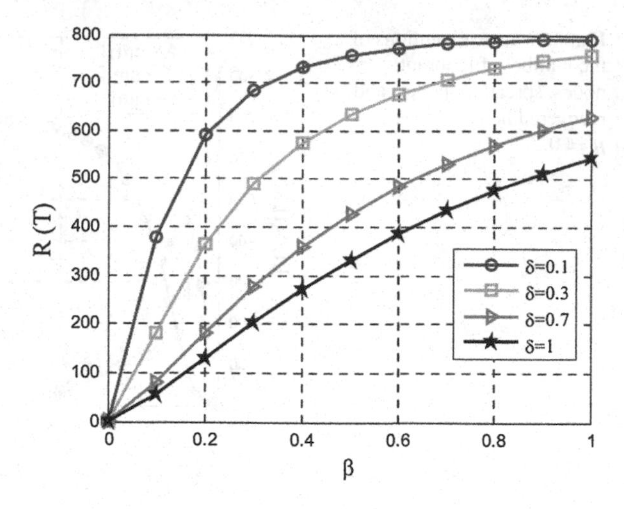

Fig. 2.6 The final number of recovered nodes R with different spreading parameter β and different immune parameter δ

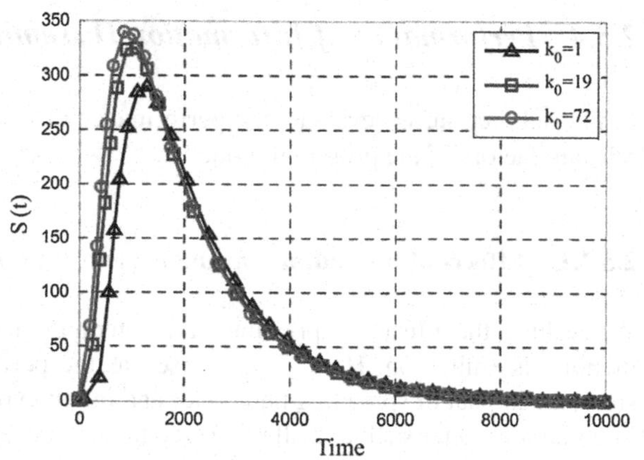

Fig. 2.7 Time evolutions of the number of spreading nodes $S(t)$ with different initial spreading nodes

2.5.4.2 Effects of Node Degree

We study the effect of the initially spreading node with different degrees. We assume that the degree of the initially spreading node is k_0. Three categories of degrees are selected for comparison to study the impact, which are $k_0 = 1$, $k_0 = 19$ and $k_0 = 72$. The result is shown by Fig. 2.7 which describes the time evolutions of the number of spreading nodes $S(t)$ with different initial spreading nodes. It is observed that the initial spreader's degree has a significant effect on information dissemination. When the degree is larger, the information dissemination speed is faster and the peak value of $S(t)$ is also larger. Actually, in MSNs, the nodes with large degree can accelerate the information dissemination, because they can recommend the information to a large number of their friends.

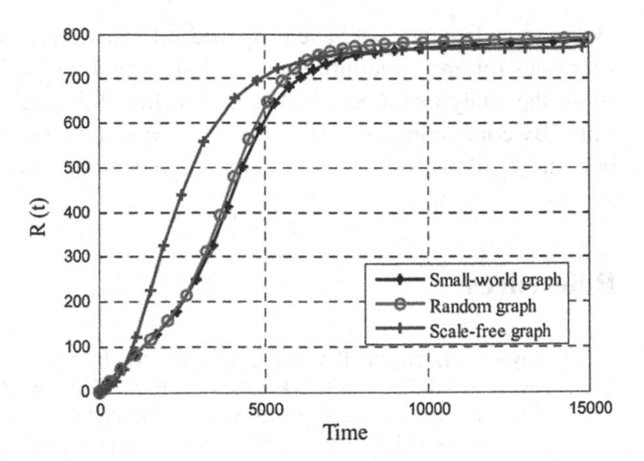

Fig. 2.8 Comparison of information dissemination based on small world graph, random graph and scale-free graph

2.5.4.3 Effects of MSN-Based Social Graph

We evaluate the performance of our model based on different types of MSN-based social graphs, which are scale-free graph model [45], random graph model [46], and small-world model [47]. All of the three networks have the same number of nodes and average degree, where $\langle k \rangle = 2$ and $|V| = 802$. From Fig. 2.8, we can see that the dissemination speed is slower in the small-world graph than that in a random graph. The reason is that the clustering coefficient in the small-world graph is larger than that in the random graph [46]. As the clustering coefficient shows how close the neighbor nodes are, there are many edges interconnected among nodes when the clustering coefficient is large. Therefore, the information is transmitted more quickly and widely in a network. Figure 2.8 also shows that information dissemination in a scale-free graph is the fastest. One reason is that the scale-free graph follows the power-law distribution. If a node with high degree becomes a spreader, a large number of nodes receive information from this spreading node at the early stage of dissemination process. Then, most nodes are spreading nodes and become recovered nodes quickly. Similarly, the recovered node with a higher degree can make spreading nodes to become recovered nodes more quickly. The other reason is that the characteristic path length of the scale-free graph is the smallest compared with others [48]. Here, the characteristic path length represents the average length of the shortest path between two nodes, randomly selected from the graphs [46]. Therefore, in scale-free graph, the information can be quickly delivered to the given node. More details can be seen in [49].

2.6 Summary

In this chapter, we have developed an analytical model to analyze the epidemic information dissemination in MSNs. The model considers the change of mobile nodes' interests by introducing two novel elements called pre-immunity and immunity.

With these elements, we have proposed information dissemination mechanism with four rules for epidemic information dissemination process. According to these four rules, the analytical model has been developed through ordinary differential equations. By conducting extensive trace-driven simulations, we have demonstrated that our analytical model is more accurate than existing ones.

References

1. N. Kayastha, D. Niyato, P. Wang, E. Hossain, Applications, architectures, and protocol design issues for mobile social networks: a survey. Proc. IEEE **99**(12), 2130–2158 (2011)
2. Y. Wang, P. Ren, F. Gao, Z. Su, A hybrid underlay/overlay transmission mode for cognitive radio networks with statistical quality-of-service provisioning. IEEE Trans. Wirel. Commun. **13**(3), 1482–1498 (2014)
3. K. Zhang, X. Liang, X. Shen, R. Lu, Exploiting multimedia services in mobile social networks from security and privacy perspectives. IEEE Commun. Mag. **52**(3), 58–65 (2014)
4. X. Liang, K. Zhang, X. Shen, X. Lin, Security and privacy in mobile social networks: challenges and solutions. IEEE Wirel. Commun. **21**(1), 33–41 (2014)
5. F. Nazir, J. Ma, A. Seneviratne, Time critical content delivery using predictable patterns in mobile social networks, in *Proceedings of the CSE*, vol. 4 (Vancouver, 2009), pp. 1066–1073
6. Y. Zhang, L. Song, W. Saad, Z. Dawy, Z. Han, Exploring social ties for enhanced device-to-device communications in wireless networks, in *Proceedings of the IEEE GLOBECOM* (Atlanta, 2013), pp. 4597–4602
7. Y. Zhang, E. Pan, L. Song, W. Saad, Z. Dawy, Z. Han, Social network aware device-to-device communication in wireless networks. IEEE Trans. Wirel. Commun. **14**(1), 177–190 (2015)
8. Z. Wang, Y. Chen, C. Li, Corman: a novel cooperative opportunistic routing scheme in mobile ad hoc networks. IEEE J. Sel. Areas Commun. **30**(2), 289–296 (2012)
9. Y. Wang, M.C. Chuah, Y. Chen, Incentive based data sharing in delay tolerant mobile networks. IEEE Trans. Wirel. Commun. **13**(1), 370–381 (2014)
10. F. Kevin, A delay-tolerant network architecture for challenged internets, in *Proceedings of the CATAPCC* (New York, 2003), pp. 27–34
11. A. Khosravi, J. Pan, Exploring personal interest in intermittently connected wireless mobile social networks, in *Proceedings of the IEEE CCNC* (Las Vegas, 2011), pp. 503–508
12. H. Zhu, S. Du, M. Li, Z. Gao, Fairness-aware and privacy-preserving friend matching protocol in mobile social networks. IEEE Trans. Emerg. Top. Comput. **1**(1), 192–200 (2013)
13. L. Buttyan, J.-P. Hubaux, *Security and cooperation in wireless networks: thwarting malicious and selfish behavior in the age of ubiquitous computing* (Cambridge University Press, Cambridge, 2007)
14. C. Zou, D. Towsley, W. Gong, Modeling and simulation study of the propagation and defense of internet e-mail worms. IEEE Trans. Dependable Secur. Comput. **4**(2), 105–118 (2007)
15. Y. Wang, X. Yang, Y. Han, X. Wang, Rumor spreading model with trust mechanism in complex social networks. Commun. Theor. Phys. **59**(4), 510–516 (2013)
16. S. Boccaletti, V. Latora, Y. Moreno, Complex networks: structure and dynamics. Phys. Rep. **424**(4), 175–308 (2006)
17. H. Wang, J. Han, L. Deng, K. Cheng, Dynamics of rumor spreading in mobile social networks. Acta Phys. Sin. **62**(11), 1–12 (2013)
18. F. Xiong, Y. Liu, Z. Zhang, Z. Jiang, Y. Zhang, An information diffusion model based on retweeting mechanism for online social media. Phys. Lett. **376**(30), 2103–2108 (2012)
19. Y. Zhang, Y. Liu, H. Zhang, H. Cheng, F. Xiong, The research of information dissemination model on online social network. Acta Phys. Sin. **60**(5), 1–7 (2011)

20. K. Zhang, X. Liang, R. Lu, X. Shen, H. Zhao, Vslp: Voronoi-socialspot-aided packet forwarding protocol with receiver location privacy in msns, in *Proceedings of the IEEE GLOBECOM* (CA, 2012), pp. 348–353
21. M. Nekovee, Y. Moreno, G. Bianconi, M. Marsili, Theory of rumour spreading in complex social networks. Phys. A **374**(1), 457–470 (2007)
22. W. Zhang, Y. Ye, H. Tan, Q. Dai, T. Li, Information diffusion model based on social network, in *Proceedings of the ICMCSA* (Wuhan, 2013), pp. 145–150
23. H. Dang, H. Wu, Clustering and cluster-based routing protocol for delay-tolerant mobile networks. IEEE Trans. Wirel. Commun. **9**(6), 1874–1881 (2010)
24. E. Talipov, Y. Chon, H. Cha, Content sharing over smartphone-based delay-tolerant networks. IEEE Trans. Mob. Comput. **12**(3), 581–595 (2013)
25. E. Sammou, Efficient probabilistic routing in delay tolerant networks, in *Proceedings of the ICMCS* (Tangier, 2012), pp. 584–589
26. Y. Ma, A. Jamalipour, A cooperative cache-based content delivery framework for intermittently connected mobile ad hoc networks. IEEE Trans. Wirel. Commun. **9**(1), 366–373 (2010)
27. K. Zhang, X. Liang, M. Barua, R. Lu, X. Shen, Phda: a priority based health data aggregation with privacy preservation for cloud assisted wbans. Inf. Sci. **284**, 1–12 (2014)
28. K. Zhang, X. Liang, R. Lu, X. Shen, Exploiting private profile matching for efficient packet forwarding in mobile social networks. Oppor. Mob. Soc. Netw. 283–312 (2014)
29. E. Bulut, B. Szymanski, Exploiting friendship relations for efficient routing in mobile social networks. IEEE Trans. Parallel Distrib. Syst. **23**(12), 2254–2265 (2012)
30. Y. Wang, J. Wu, W. Yang, Cloud-based multicasting with feedback in mobile social networks. IEEE Trans. Wirel. Commun. **12**(12), 6043–6053 (2013)
31. P. Costa, C. Mascolo, M. Musolesi, G. Picco, Socially-aware routing for publish-subscribe in delay-tolerant mobile ad hoc networks. IEEE J. Sel. Areas Commun. **26**(5), 748–760 (2008)
32. H. Sun, C. Wu, Epidemic forwarding in mobile social networks, in *Proceedings of the IEEE ICC* (Ottawa, 2012), pp. 1421–1425
33. Y. Wu, S. Deng, H. Huang, Hop limited epidemic-like information spreading in mobile social networks with selfish nodes. J. Phys. A: Math. Theor. **46**(26), 1–14 (2013)
34. D. Adu-Gyamfi, Y. Wang, F. Zhang, M. Domenic, I. Memon, Y. Gustav, Modeling the spreading behavior of passive worms in mobile social networks, in *Proceeding of the ICIII*, vol. 1 (Xi'an, 2013), pp. 380–383
35. Y. Wu, S. Deng, H. Huang, Information propagation through opportunistic communication in mobile social networks. Mob. Netw. Appl. **17**(6), 773–781 (2012)
36. K. Zhang, X. Liang, R. Lu, X. Shen, Pif: a personalized fine-grained spam filtering scheme with privacy preservation in mobile social networks. IEEE Trans. Comput. Soc. Syst. **2**(3), 41–52 (2015)
37. A. Clauset, C. Shalizi, M. Newman, Power-law distributions in empirical data. SIAM Rev. **51**(4), 661–703 (2009)
38. A. Clementi, C. Macci, A. Monti, F. Pasquale, R. Silvestri, Flooding time in edge-markovian dynamic graphs, in *Proceedings of the ACM PDC* (Toronto, 2008), pp. 213–222
39. M. Karaliopoulos, Assessing the vulnerability of dtn data relaying schemes to node selfishness. IEEE Commun. Lett. **13**(12), 923–925 (2009)
40. Y. Li, G. Su, D. Wu, D. Jin, L. Su, L. Zeng, The impact of node selfishness on multicasting in delay tolerant networks. IEEE Trans. Veh. Technol. **60**(5), 2224–2238 (2011)
41. H. Zhu, L. Fu, G. Xue, Y. Zhu, M. Li, L. Ni, Recognizing exponential inter-contact time in vanets, in *Proceedings of the IEEE INFOCOM* (San Diego, 2010), pp. 1–5
42. T. Karagiannis, J.L. Boudec, M. Vojnovic, Power law and exponential decay of intercontact times between mobile devices. IEEE Trans. Mob. Comput. **9**(10), 1377–1390 (2010)
43. R. Boas, Counterexamples to l'hopital's rule. Am. Math. Mon. **93**(9), 644–645 (1986)
44. A. Vázquez, M. Weigt, Computational complexity arising from degree correlations in networks. Phys. Rev. E **67**(2), 1–4 (2003)
45. R. Albert, A.-L. Barabási, Statistical mechanics of complex networks. Rev. Mod. Phys. **74**(1), 47–49 (2002)

46. M. Newman, S. Strogatz, D. Watts, Random graphs with arbitrary degree distributions and their applications. Phys. Rev. E **64**(2), 1–17 (2001)
47. D. Watts, S. Strogatz, Collective dynamics of small-worldnetworks. Nature **393**(6684), 440–442 (1998)
48. B. Tian, D. Towsley, On distinguishing between internet power law topology generators, in *Proceedings of the IEEE INFOCOM* (Orlando, 2002), pp. 638–647
49. Q. Xu, Z. Su, K. Zhang, P. Ren, X. Shen, Epidemic information dissemination in mobile social networks with opportunistic links. IEEE Trans. Emerg. Top. Comput. **3**(3), 399–409 (2015)

Chapter 3
Modeling of Selfishness-Aware Incentive for MSNs

In this chapter, we discuss the modeling of selfishness and propose an incentive mechanism in MSNs. We consider two types of selfish behaviors: weak selfishness and extreme selfishness. The information dissemination with weak and extreme selfishness are studied, respectively, through the ordinary differential equations (ODEs). Then, an incentive model is developed to stimulate users to participant cooperation.

3.1 Selfishness-Aware Incentive in MSNs

Due to the rapid growth of wireless communication [1, 2] and sensor technologies [3], various kinds of content [4–6] can be provided to both mobile networks and users. Currently, mobile crowd sensing (MCS) [7] has emerged as a new paradigm of MSNs. A large number of mobile individuals can use their mobile devices to deliver data, which can be provided to the MCS to collect the information of interest [8–13]. Based on the collected information [14, 15] from the participants, the MCS can get knowledge or make decision for the large scale systems. For example, the passengers on the bus can use their mobile phones to send the real time information, including the current location, the status of traffic, and the situation of overcrowding. Then, with the above information, the MCS can sense the environment around the bus and predict the arriving time, the number of spared seats on the bus, etc. However, it needs all participants to cooperatively collect the information from individuals. How to motivate participation of mobile individuals is an important issue.

However, in practice a large number of users exhibit various degrees of selfishness [16]. For example, some users mainly concern how to save their limited resources (e.g., battery, buffer, etc.), and thus they are not willing to provide information to others. Obviously, users' selfishness will affect the performance of mobile crowds. In the previous studies, the selfishness is mainly divided into two parts: individual selfishness and social selfishness. The first is from the perspective of an individual, in which the user is not willing to relay and store information for others to save the limited buffer and power resources [17]. The second is from the perspective of a

Z. Su et al., *Modeling and Optimization for Mobile Social Networks*,
DOI 10.1007/978-3-319-47922-4_3

community formed by some individuals with similar interests, where the individuals are more willing to help others in the same community but lack interest to contribute information to the ones out of the community [18].

In this chapter, different from the above conventional work, we characterize users' selfishness based on different levels: weak selfishness and extreme selfishness. The weak selfishness means that an individual may be unwilling to forward information to others in order to conserve the limited resources but sometimes still has interest to contribute information to others. The extreme selfishness refers that an individual does not want to share information to help others. Here, the extreme selfishness is considered from two aspects. On one hand, some users who are not interested in the information will not accept or store it. On the other hand, some users who have possessed the information have no will to forward it to others. Then, an analytical model for information dissemination with weak and extreme selfishness is presented through the ordinary differential equations (ODEs). This model can be employed to evaluate the influences of both the extreme selfish and weak selfish behavior on the process of dissemination. In addition, real trace driven simulations are conducted to show the effect of both weak selfishness and extreme selfishness in information dissemination.

3.2 Related Work

Recently, crowd sensing [7, 19, 20] has attracted a lot of attentions. Bulut et al. [21] introduce a friendship-based routing scheme, where a novel metric is used to accurately detect the quality of friendship and make the forwarding decisions. Lee et al. [22] present the task distribution models and improve the efficiency of data sensing. Sheng et al. [23] design a probabilistic mode and discuss the scheduling methods, when the GPS can not provide accurate location information for sensing coverage. Guo et al. [24] use the passive interference power received on mobile devices to sense the volume of wireless throughput of users, which is non-intrusive way without requiring data from private devices. An et al. [25] consider the credible interaction issues between users and introduce a crowdsourcing assignment method based on the users' social relationship cognition to make sure the crowdsourcing tasks to be assigned to credible users. Guo et al. [19] present a group-aware mobile crowd sensing system which can facilitate group formation and management by using users' online and offline social behaviors. However, these methods have not discussed how to encourage mobile individuals to contribute data during information dissemination.

There have been many studies on evaluating the performance of the information dissemination in mobile networks. Talasila et al. [20] provide a scheme to improve the location reliability of mobile crowd sensed data, where the trust is bootstrapped in the system with image processing techniques validating a fewer photos submitted by users. Wen et al. [26] present a quality-driven based incentive scheme for the crowd sensing system, where the participant is rewarded based on the quality of sensed data instead of working data. Sun et al. [27] propose a model for unicast epidemic message forwarding in the MSN, which considers the message validity. Wu et al. [28] present a theoretical model to evaluate the impact of people's behaviors on information propagation. Wang et al. [29] study the biggest speed and distance that the data packed can have when it is disseminated in a mobile opportunistic network. Both the one-copy case and the multiple-copy case have been studied in the small scale and lager scale mobile network. Fan et al. [30] propose a data broadcasting method in the MSN which uses the users' movement and community information to decide the broadcast routes. However, all these models have not considered the selfishness to explore the social nature of the networks in the information dissemination.

As for the node selfishness, Mei et al. [31] introduce two forwarding protocols for mobile wireless network based on the features of selfish individuals. Li et al. [18] present a routing performance by using a social selfishness-aware routing algorithm in delay tolerant networks. Li et al. [32] develop a model to investigate the impact of selfish behaviors on the performance of multicast in delay tolerant networks through a 3-D continuous time Markov chain. Hernndez-Orallo et al. [33] introduce a collaborative watchdog mechanism with fast diffusion of selfish nodes awareness to detect selfish nodes and present a model to evaluate the cost of the detection. Choi et al. [34] study the impact of users' selfish for the data replica allocation in MSNs and develop a selfish node detection algorithm. Li et al. [35] study the node's selfishness in the mobile network and divide the selfishness into three classes which are node-selfishness, intrinsic selfishness, and extrinsic selfishness. Although the selfishness has been studied extensively, they are not specially designed for mobile social networks or crowd sensing.

3.3 System Model

This section introduces a system model which includes social network, mobility model, and node selfishness.

3.3.1 Social Network

We study the crowd sensing with mobile nodes in the environment of mobile social networks, where the system structure is shown by Fig. 3.1. It is assumed that there are N users denoted by $n_1, n_2, \ldots, n_N$ moving around within a region all the time. The information can be shared between two users only when they meet and have social relations.

The social network consists of opportunistic network and social network. It is modeled by an unweighted and undirected graph $G(V, E)$. The symbol V is defined as the set of nodes, and E denotes the set of edges. The nodes represent users, and their social ties (e.g., friendship and relatives) are denoted by the edges. Thus, we

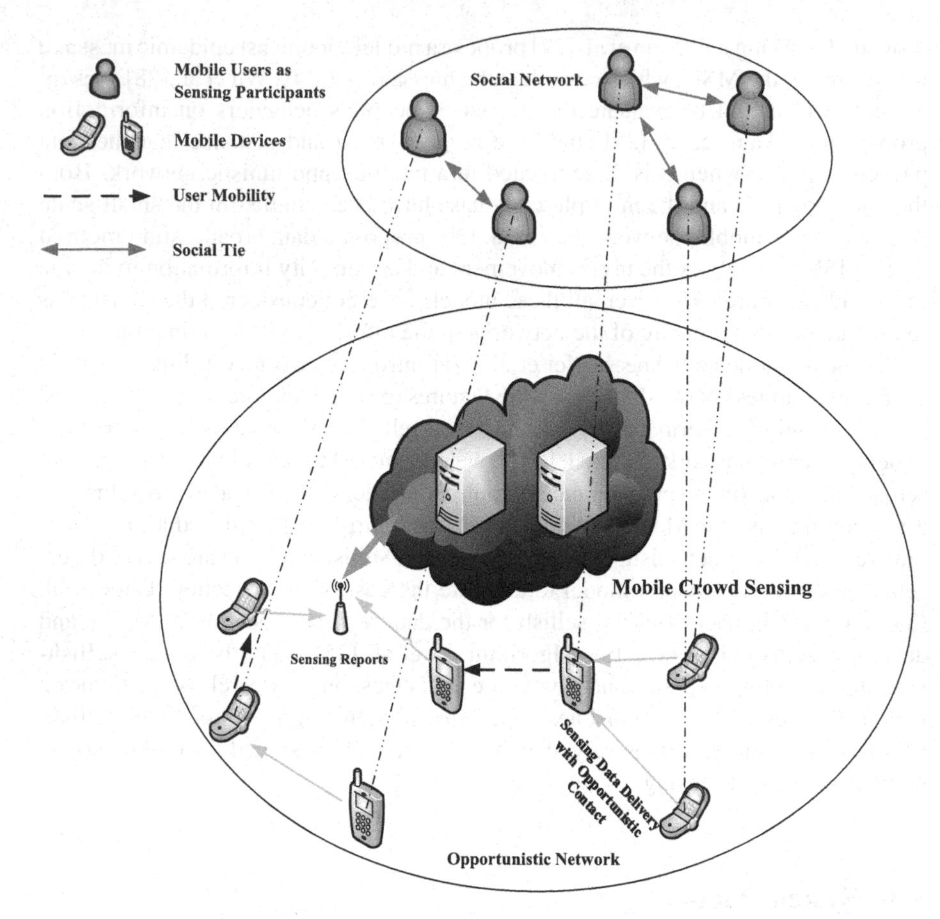

Fig. 3.1 System structure

have $V = \{n_1, n_2, \ldots, n_N\}$. To study the impact of selfishness, the distribution of relationship should be acquired. The probability of a node having k friends is defined by $P(k)$, where the degree of the node is k. Related works show that $P(k)$ follows the power-low distribution in the mobile social network [36]. Therefore, $P(k)$ [27] can be obtained as follows:

$$P(k) = \begin{cases} 0, & k < m \\ C(m, \gamma)k^{-\gamma}, & m \leq k < N \end{cases}. \tag{3.1}$$

Here, the smallest degree of the network is m, the skewness of the degree distribution is γ, and the normalization constant is $C(m, \gamma)$.

3.3.2 Mobility Model

As the mobility patterns of nodes are sophisticated, most of models with high accuracy to capture the mobility patterns are not applicable because of the computational complexity. At present, some simple mobility models are still used in some works, such as [17, 27]. These simple models typically have an exponentially distributed inter-meet time, which has been demonstrated with real trace by some studies [37, 38]. Therefore, in this chapter, it is assumed that the meeting time epochs of each node conforms to a Poisson distribution with parameter λ, giving rise to exponentially distributed inter-meet time between two nodes.

3.3.3 Node Selfishness

The concept of selfishness denotes that mobile nodes are not willing to help their friends to forward information due to the limited resources. Generally, mobile nodes have two patterns to exhibit selfishness. The first one is that mobile nodes occasionally contribute information to others. The second one is that some nodes do not accept or contribute information at all. Therefore, the selfishness can be divided into extreme selfishness and weak selfishness. Specially, it is assumed that (1) a mobile node does not forward information to one of its friends with a probability p_{nf}, which denotes the weak selfishness level; (2) a mobile node does not accept the information from others absolutely, where the corresponding parameter is p_1 with an exponential distribution; (3) a mobile node refuses to forward the information to any nodes, where the corresponding parameter is p_2 with an exponential distribution. Here, the first one refers to the weak selfishness and the next two assumptions are related to the extreme selfishness.

3.4 Proposed Model with a Novel Selfishness Division

It is assumed that there is only one node having information initially, and then this node moves randomly in a finite region. When this node encounters its friends who do not have the information, the information can be delivered.

$I_k(t)$ is denoted as the number of k -degree nodes which do not have information at time t. $S_k(t)$ is defined as the number of k -degree nodes which obtain information at time t. $R_k(t)$ means the number of k -degree nodes which refuse accepting the information at time t. $Z_k(t)$ is defined as the number of k-degree nodes which have received the information but don't want to forward it to the next node at time t. Let $N_k(t)$ denote the number of k -degree nodes at a given time t. Then, the fractions of these four types of nodes in $N_k(t)$ becomes as follows at time t.

$$\begin{cases} i_k(t) = I_k(t)/N_k(t) \\ s_k(t) = S_k(t)/N_k(t) \\ r_k(t) = R_k(t)/N_k(t) \\ z_k(t) = Z_k(t)/N_k(t) \end{cases}. \tag{3.2}$$

Within the interval $[t, t + \Delta t]$, we can have the variation of $i_k(t)$ by

$$i_k(t + \Delta t) = i_k(t)(1 - P(G, k)). \tag{3.3}$$

Here $P(G, k)$ is defined as the probability that a k-degree node n_i without information receives it or refuses accepting it from any one within Δt.

We consider two aspects to derive $P(G, k)$. On one hand, node n_i receiving the information from other friends is classified by $s_k(t)$. On the other hand, node n_i which may refuse accepting information is classified by $r_k(t)$.

As mentioned above, only when an opportunistic link exists between two nodes, they are able to have a chance to exchange information. In addition, since mobile nodes forward information only to their friends, the social tie between two nodes should also exist. As the inter-meet between two nodes follows an exponential distribution, the probability that node n_i encounters other nodes within the interval of $[t, t + \Delta t]$ becomes $1 - e^{-\lambda(N-1)\Delta t}$. In fact, if the interval is short enough, node n_i can only encounter one node. Thus, probability that the node n_i meets the other node (e.g., n_j.) within $[t, t + \Delta t]$ is $1 - e^{-\lambda(N-1)\Delta t}$. Since the degree of node n_i is k, the probability that node n_j is a friend of n_i can be obtained by $k/(N - 1)$. Besides, the probability that node n_j has information is $\sum_{k'=m}^{N-1} \frac{N}{N-1} P(k') s_{k'}(t)$, where the degree of a node is the k'. Then, node n_i may get information successfully from one of its friends when these two nodes encounter with the probability $1 - p_{nf}$.

Therefore, according to the above description, the probability $P(A, k)$ that node n_i can receive the information from its friend within Δt can be obtained by

$$P(A, k) = (1 - p_{nf}) \frac{k}{N - 1} (1 - e^{-\lambda(N-1)\Delta t}) \sum_{k'=m}^{N-1} \frac{N}{N - 1} P(k') s_{k'}(t). \tag{3.4}$$

Here, the above shows the necessary conditions that node n_i can receive information. Firstly, node n_i should encounter node n_j, and this node n_j is a friend of node n_i. Secondly, node n_j should have information and be willing to forward information to node n_i.

Then, about the second aspect, the probability $P(not\ receive)$ that node n_i does not accept information absolutely within Δt becomes

$$P(not\ receive) = \int_0^{\Delta t} p_1 e^{-p_1 t} dt = 1 - e^{-p_1 \Delta t}. \tag{3.5}$$

By combining with (3.3) and (3.5), we obtain

$$
\begin{aligned}
P(G,k) &= 1-(1-P(A,k))(1-P(not\,receive)) \\
&= 1-\left(1-(1-p_{nf})\left(\frac{k}{N-1}\right)\left(1-e^{-\lambda(N-1)\Delta t}\right)\right. \\
&\quad \left.\times\sum_{k'=m}^{N-1}\frac{N}{N-1}P(k')s_{k'}(t)\right)\times e^{-p_1\Delta t}.
\end{aligned}
\tag{3.6}
$$

From (3.3), we have

$$
\begin{aligned}
\frac{di_k(t)}{dt} &= -i_k(t)\lim_{\Delta t\to 0}\frac{P(G,k)}{\Delta t} \\
&= -i_k(t)\lim_{\Delta t\to 0}\frac{1-\left(1-(1-p_{nf})(k/N-1)\left(1-e^{-\lambda(N-1)\Delta t}\right)\right)}{\Delta t} \\
&\quad\times\sum_{k'=m}^{N-1}\frac{N}{N-1}P(k')s_{k'}(t)e^{-p_1\Delta t} \\
&= -\lambda k(1-p_{nf})i_k(t)\sum_{k'=m}^{N-1}\frac{N}{N-1}P(k')s_{k'}(t)-p_1 i_k(t).
\end{aligned}
\tag{3.7}
$$

Similarly, for a short time interval Δt, it can be obtained by,

$$
s_k(t+\Delta t)=s_k(t)+i_k(t)P(A,k)-s_k(t)P(not\,forward). \tag{3.8}
$$

where $P(not\,forward)$ denotes the probability that a k -degree node having the information refuses to forward it to any of other nodes within $[t, t+\Delta t]$, which can be computed as follows:

$$
P(not\,forward)=\int_0^{\Delta t}p_2 e^{-p_2\Delta t}dt = 1-e^{-p_2\Delta t}. \tag{3.9}
$$

We have

$$
\begin{aligned}
\frac{ds_k(t)}{dt} &= \lim_{\Delta t\to 0}\frac{i_k(t)P(A,k)-s_k(t)P(not\,forward)}{\Delta t} \\
&= \lim_{\Delta t\to 0}\frac{i_k(t)\left(\frac{(1-p_{nf})k}{N-1}\left(1-e^{-\lambda(N-1)\Delta t}\right)\sum_{k'=m}^{N-1}\frac{N}{N-1}P(k')s_{k'}(t)\right)}{\Delta t} \\
&\quad -\frac{s_k(t)\left(1-e^{-p_2\Delta t}\right)}{\Delta t} \\
&= \lambda k(1-p_{nf})i_k(t)\sum_{k'=m}^{N-1}\frac{N}{N-1}P(k')s_{k'}(t)-p_2 s_k(t).
\end{aligned}
\tag{3.10}
$$

In addition, given a short time interval Δt, we can derive

$$r_k(t+\Delta t) = r_k(t) + i_k(t)P(not\ receive). \tag{3.11}$$

Therefore, we have

$$\begin{aligned}\frac{dr_k(t)}{dt} &= \lim_{\Delta t \to 0} \frac{i_k(t)P(not\ receive)}{\Delta t} \\ &= i_k(t) \lim_{\Delta t \to 0} \frac{1-e^{-p_1 \Delta t}}{\Delta t} \\ &= p_1 i_k(t).\end{aligned} \tag{3.12}$$

Given a short time interval Δt, we have

$$z_k(t+\Delta t) = z_k(t) + s_k(t)P(notforward). \tag{3.13}$$

Then, we have

$$\begin{aligned}\frac{dz_k(t)}{dt} &= \lim_{\Delta t \to 0} \frac{s_k(t)P(notforward)}{\Delta t} \\ &= s_k(t) \lim_{\Delta t \to 0} \frac{1-e^{-p_2 \Delta t}}{\Delta t} \\ &= p_2 s_k(t).\end{aligned} \tag{3.14}$$

Therefore, the ordinary differential equations (ODEs) to model the dynamic information dissemination among the mobile individuals are shown as follows:

$$\frac{di_k(t)}{dt} = -\lambda(1-p_{nf})ki_k(t)\sum_{k'=m}^{N-1} \frac{N}{N-1}P(k')s_{k'}(t) - p_1 i_k(t). \tag{3.15}$$

$$\frac{ds_k(t)}{dt} = \lambda k(1-p_{nf})i_k(t)\sum_{k'}^{N-1} \frac{N}{N-1}P(k')s_{k'}(t) - p_2 s_k(t). \tag{3.16}$$

$$\frac{dr_k(t)}{dt} = p_1 i_k(t). \tag{3.17}$$

$$\frac{dz_k(t)}{dt} = p_2 s_k(t). \tag{3.18}$$

3.5 Performance Evaluation

Based on the derived analytical mode, this section is to evaluate the effect of the weak and extreme selfishness on the information dissemination in MSNs.

3.5.1 Simulation Setup

It is assumed that the unit time of the system is 0.01 h. Since related studies show that the average inter-meet time between two nodes is around 5 h [27, 37], λ is set as 0.002. In addition, the considered MSN is constructed by the real trace from logs of a mobile social system. Two people are friends if they have at least one record in the trace. Let $\gamma = 2.1$, which is a typical exponent of a scale-free network. Basic parameters of the social graph are $N = 802$, $E = 1222$, and $m = 1$, respectively.

In the simulation, the number of infected nodes $F(t)$ is used as the performance metrics. $F(t)$ denotes the average number of nodes that have received information at time t. It also represents the efficiency of an information dissemination scheme in the network, in terms of how widely a message is propagated.

3.5.2 Effect of Weak Selfishness

Here, the effect of weak selfishness on the information dissemination is evaluated with numerical results by the proposed model.

The settings of the analytical model are as follows. $p_1 = p_2 = 0$. p_{nf} varies from 0 to 1 while $T = 1000, 2000, 5000, 10{,}000$, respectively.

Figure 3.2 shows the effect of weak selfishness. It is observed that the weak selfishness is negatively related with the number of nodes having information, as $F(T)$ decreases with the growth of weak selfishness level p_{nf} in each curve. In addition, it can be seen that the network is much more robust to the weak selflessness when the dissemination time is larger. For instance, in the case of $T = 10{,}000$, $F(T)$ is approximately stable when p_{nf} is smaller than 0.7. Oppositely, when $T = 1000$ or $T = 2000$, $F(T)$ decreases rapidly with the weak selflessness level increasing.

We also investigate the effect of weak selfishness under different mobility models with different inter-meet intensity $\lambda = 0.101, 0.084, 0.051$ [39], which are obtained based on statistics of the Cambridge trace dataset [40]. Here, $p_1 = p_2 = 0, T = 200$. p_{nf} is increased from 0 to 0.1. The effects of weak selfishness with different inter-meet intensities are shown in Fig. 3.3. When λ is larger, the effect of the weak selfishness is smaller. In fact, if λ is larger, the probability of encounter between any node pair is larger, resulting in a better information dissemination.

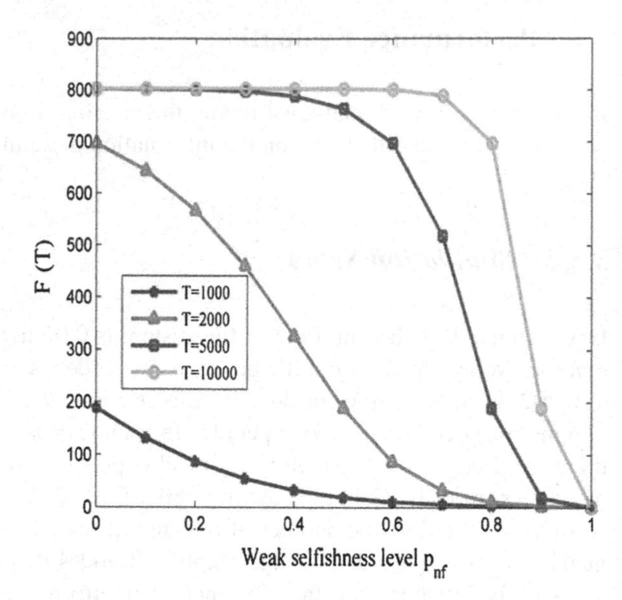

Fig. 3.2 Effect of weak selfishness

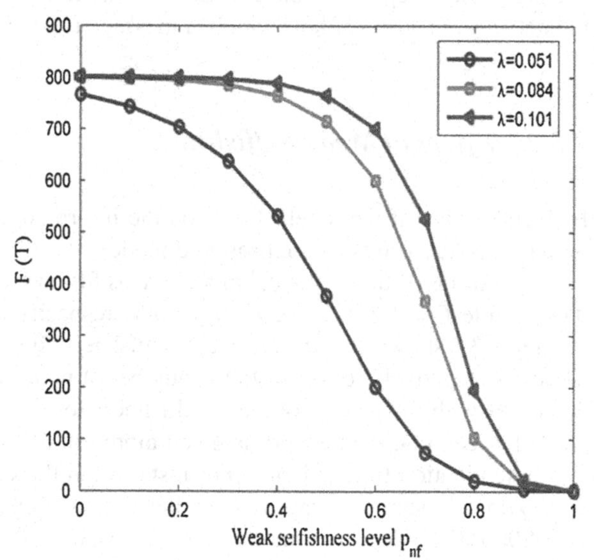

Fig. 3.3 Effect of weak selfishness when the λ is different

3.5.3 Effect of Extreme Selfishness

Here, the effect of extreme selfishness on the information dissemination is evaluated with numerical results by the presented model.

As mentioned above, the extreme selfishness is divided into two aspects, where the first aspect is related to p_1 and the second aspect is related to p_2. Thus, to evaluate the effect of the first aspect of extreme selfishness, five types of p_1 is chosen for

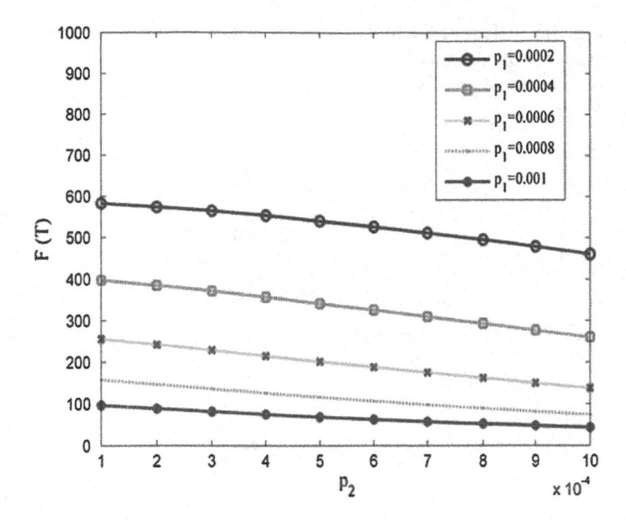

Fig. 3.4 Effect of extreme selfishness

comparison. Let p_2 vary from 0.0001 to 0.001 to investigate the effect the second aspect of extreme selfishness. Here, $p_{nf} = 0$ and $T = 5000$. Based on these settings, the effect of extreme selfishness is shown in Fig. 3.4.

From Fig. 3.4, it can be observed that the extreme selfishness restrains the dissemination of information. Firstly, when p_2 is fixed (e.g., $p_2 = 5 \times 10^{-4}$), the value of $F(T)$ gradually decreases with the increase of the value of p_1. Thus, the first aspect of extreme selfishness can hinder the information dissemination. Secondly, when the level of the first aspect of extreme selfishness is constant (e.g., $p_1 = 0.0006$), the value of $F(T)$ also decreases with the increase of the value of p_2, which means that the second aspect of the extreme selfishness also has a negative effect on the information dissemination.

Next, the relationship between the extreme selfishness and λ is investigated. Firstly, the experiment on the first aspect of the extreme selfishness is carried out, where $p_2 = 0$, $p_{in} = 1$, and p_1 increases from 0 to 0.005. Then, experiment on the second aspect of the extreme selfishness is implemented, where $p_1 = 0$, $T = 100$, and p_2 increases from 0 to 0.005.

Figure 3.5 shows the effect of the first aspect for extreme selfishness with different λ. It indicates that all the three curves descend almost linearly with p_1, but the curve of smaller λ is much more sensitive to the variation of p_1. Therefore, if λ is smaller, the effect of the first aspect of extreme selfishness is larger. Figure 3.6 shows the effect of the second aspect for extreme selfishness with different λ, which illustrates the relationship between the second aspect of extreme selfishness and λ. When λ is larger, the network is much more robust to the weak selfishness. Accordingly, if λ is smaller, the effect of the second aspect of extreme selfishness is more significant. More details can be seen in [41].

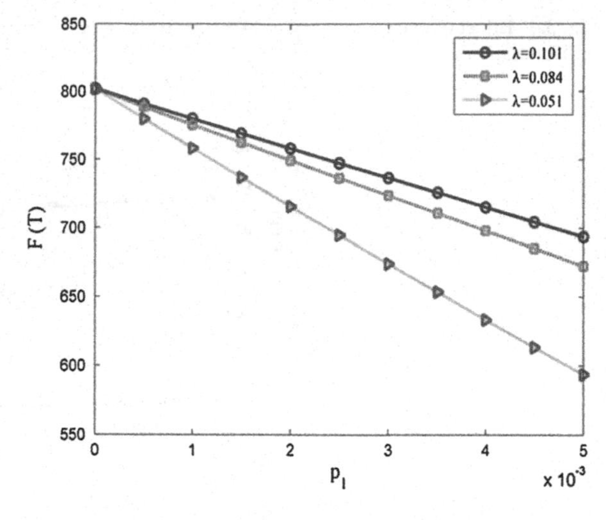

Fig. 3.5 Effect of the first aspect of the extreme selfishness when λ is different

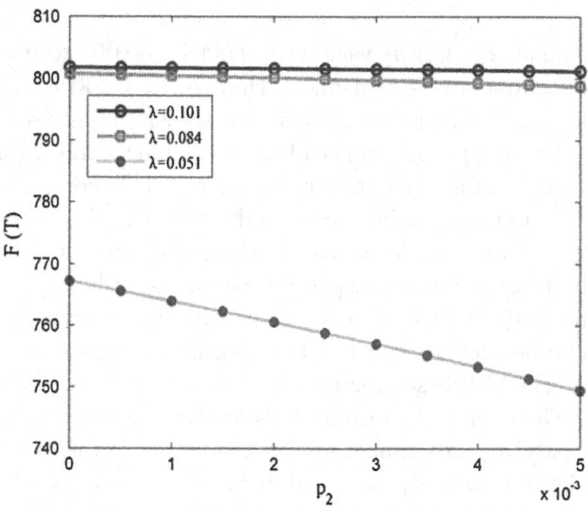

Fig. 3.6 Effect of the second aspect of the extreme selfishness when λ is different

3.6 Summary

This chapter has presented an analytical model for modeling and understanding the effect of selfishness, in order to encourage the individuals to participant the cooperation of crowd sensing in mobile social networks. The proposed analytical model has considered two new types of selfishness: weak selfishness and extreme selfishness. The simulations based on the real trace have shown the effects of weak selfishness and extreme selfishness.

References

1. J. Andrews, S. Buzzi, W. Choi, S. Hanly, A. Lozano, A. Soong, J. Zhang, What will 5G be? IEEE J. Sel. Areas Commun. **32**(6), 1065–1082 (2014)
2. H. Zhu, S. Du, Z. Gao, M. Dong, Z. Cao, A probabilistic misbehavior detection scheme toward efficient trust establishment in delay-tolerant networks. IEEE Trans. Parallel Distrib. Syst. **25**(1), 22–32 (2014)
3. R. Du, C. Chen, B. Yang, N. Lu, X. Guan, X. Shen, Effective urban traffic monitoring by vehicular sensor networks. IEEE Trans. Veh. Technol. **64**(1), 273–286 (2015)
4. Z. Su, M. Oguro, Y. Okada, J. Katto, S. Okubo, Overlay tree construction to distribute layered streaming by application layer multicast. IEEE Trans. Consum. Electron. **56**(3), 1957–1962 (2010)
5. Y. Wang, P. Ren, F. Gao, Z. Su, A hybrid underlay/overlay transmission mode for cognitive radio networks with statistical quality-of-service provisioning. IEEE Trans. Wirel. Commun. **13**(3), 1482–1498 (2014)
6. T. Yang, Z. Zheng, H. Liang, R. Deng, N. Cheng, X. Shen, Green energy and content-aware data transmissions in maritime wireless communication networks. IEEE Trans. Intell. Transp. Syst. **16**(2), 751–762 (2015)
7. J. Sun, An incentive scheme based on heterogeneous belief values for crowd sensing in mobile social networks, in *Proceedings of IEEE GLOBECOM* (2013), pp. 1717–1722
8. Q. Xu, Z. Su, B. Han, D. Fang, Z. Xu, Analytical model for epidemic information dissemination in mobile social networks with a novel selfishness division, in *Proceedings of LIMS* (Shanghai, 2014), pp. 478–484
9. K. Ota, M. Dong, J. Wang, S. Guo, Z. Cheng, M. Guo, Dynamic itinerary planning for mobile agents with a content-specific approach in wireless sensor networks, in *Proceedings of IEEE VTC* (Ottawa, 2010), pp. 1–5
10. T. Yang, H. Liang, N. Cheng, R. Deng, X. Shen, Efficient scheduling for video transmissions in maritime wireless communication networks. IEEE Trans. Veh. Technol. **64**(9), 4215–4229 (2015)
11. M. Dong, T. Kimata, K. Sugiura, K. Zettsu, Quality-of-experience (QoE) in emerging mobile social networks. IEICE Trans. Inf. Syst. **97**(10), 2606–2612 (2014)
12. D. Fang, Z. Su, Q. Xu, Analysis of data transmission based on the priority over grid structures. ICIC Express Lett. Part B Appl. Int. J. Res. Surv. **5**(3), 751–755 (2014)
13. H. Zhu, X. Lin, R. Lu, Y. Fan, X. Shen, Smart: a secure multilayer credit-based incentive scheme for delay-tolerant networks. IEEE Trans. Veh. Technol. **58**(8), 4628–4639 (2009)
14. S. Zhu, L. Xie, C. Chen, X. Guan, Collective behavior of mobile agents with state-dependent interactions. Automatica **51**, 394–401 (2015)
15. Q. Xu, Z. Su, K. Zhang, P. Ren, X. Shen, Epidemic information dissemination in mobile social networks with opportunistic links. IEEE Trans. Emerg. Top. Comput. **3**(3), 399–409 (2015)
16. P. Hui, K. Xu, V. Li, J. Crowcroft, V. Latora, P. Lio, Selfishness, altruism and message spreading in mobile social networks, in *Proceedings of IEEE INFOCOM* (Rio de Janeiro, 2009), pp. 1–6
17. M. Karaliopoulos, Assessing the vulnerability of dtn data relaying schemes to node selfishness. IEEE Commun. Lett. **13**(12), 923–925 (2009)
18. Q. Li, S. Zhu, G. Cao, Routing in socially selfish delay tolerant networks, in *Proceedings of IEEE INFOCOM* (San Diego, 2010), pp. 1–9
19. B. Guo, Z. Yu, L. Chen, X. Zhou, X. Ma, MobiGroup: enabling lifecycle support to social activity organization and suggestion with mobile crowd sensing. IEEE Trans. Hum. Mach. Syst. **46**(3), 390–402 (2016)
20. M. Talasila, R. Curtmola, C. Borcea, Improving location reliability in crowd sensed data with minimal efforts, in *Proceedings of WMNC* (Dubai, 2013), pp. 1–8
21. E. Bulut, B. Szymanski, Friendship based routing in delay tolerant mobile social networks, in *Proceedings of IEEE GLOBECOM* (Miami, 2010), pp. 1–5

22. L. Pournajaf, A. Garcia-Ulloa, L. Xiong, V. Sunderam, Participant privacy in mobile crowd sensing task management: a survey of methods and challenges, in *Proceedings of ACM SIGMOD* (New York, 2014)
23. X. Sheng, J. Tang, X. Xiao, G. Xue, Leveraging GPS-less sensing scheduling for green mobile crowd sensing. IEEE Internet Things J. **1**(4), 328–336 (2014)
24. W. Guo, S. Wang, Mobile crowd-sensing wireless activity with measured interference power. IEEE Wirel. Commun. Lett. **2**(5), 539–542 (2013)
25. J. An, X. Gui, Z. Wang, J. Yang, X. He, A crowdsourcing assignment model based on mobile crowd sensing in the internet of things. IEEE Internet Things J. **2**(5), 358–369 (2015)
26. Y. Wen, J. Shi, Q. Zhang, X. Tian, Z. Huang, H. Yu, Y. Cheng, X. Shen, Quality-driven auction-based incentive mechanism for mobile crowd sensing. IEEE Trans. Veh. Technol. **64**(9), 4203–4214 (2015)
27. H. Sun, C. Wu, Epidemic forwarding in mobile social networks, in *Proceedings of IEEE ICC* (Ottawa, 2012), pp. 1421–1425
28. Y. Wu, S. Deng, H. Huang, Information propagation through opportunistic communication in mobile social networks. Mob. Netw. Appl. **17**(6), 773–781 (2012)
29. S. Wang, X. Wang, J. Huang, R. Bie, Z. Tian, F. Zhao, The potential of mobile opportunistic networks for data disseminations. IEEE Trans. Veh. Technol. **65**(2), 912–922 (2016)
30. J. Fan, J. Chen, Y. Du, W. Gao, J. Wu, Y. Sun, Geocommunity-based broadcasting for data dissemination in mobile social networks. IEEE Trans. Parallel Distrib. Syst. **24**(4), 734–743 (2013)
31. A. Mei, J. Stefa, Give2Get: forwarding in social mobile wireless networks of selfish individuals. IEEE Trans. Dependable Secure Comput. **9**(4), 569–582 (2012)
32. Y. Li, G. Su, D. Wu, D. Jin, L. Su, L. Zeng, The impact of node selfishness on multicasting in delay tolerant networks. IEEE Trans. Veh. Technol. **60**(5), 2224–2238 (2011)
33. E. Hernández-Orallo, M. Olmos, J. Cano, C. Calafate, P. Manzoni, A fast model for evaluating the detection of selfish nodes using a collaborative approach in manets. Wirel. Pers. Commun. **74**(3), 1099–1116 (2014)
34. J. Choi, K. Shim, S. Lee, K. Wu, Handling selfishness in replica allocation over a mobile ad hoc network. IEEE Trans. Mob. Comput. **11**(2), 278–291 (2012)
35. J. Li, Q. Yang, P. Gong, K. Kwak, End-to-end multiservice delivery in selfish wireless networks under distributed node-selfishness management. IEEE Trans. Commun. **64**(3), 1132–1142 (2016)
36. A. Clauset, C. Shalizi, M. Newman, Power-law distributions in empirical data. SIAM Rev. **51**(4), 661–703 (2009)
37. T. Karagiannis, J.L. Boudec, M. Vojnovic, Power law and exponential decay of intercontact times between mobile devices. IEEE Trans. Mob. Comput. **9**(10), 1377–1390 (2010)
38. H. Zhu, L. Fu, G. Xue, Y. Zhu, M. Li, L. Ni, Recognizing exponential inter-contact time in vanets, in *Proceedings of IEEE INFOCOM* (San Diego, 2010), pp. 1–5
39. J. Liu, X. Jiang, H. Nishiyama, N. Kato, A general model for store-carry-forward routing schemes with multicast in delay tolerant networks, in *Proceedings of CHINACOM* (Harbin, 2011), pp. 494–500
40. P. Hui, J. Crowcroft, E. Yoneki, Bubble rap: social-based forwarding in delay-tolerant networks. IEEE Trans. Mob. Comput. **10**(11), 1576–1589 (2011)
41. Q. Xu, Z. Su, B. Han, D. Fang, Z. Xu, X. Gan, Analytical model with a novel selfishness division of mobile nodes to participate cooperation. Peer-to-Peer Netw. Appl. **9**(4), 712–720 (2016)

Chapter 4
Optimal Relay Services for MSNs

In this chapter, the optimization of relay service for MSNs is studied. In MSNs, each user has his own virtual currency and can earn currency as a relay for other users. A bundle carrier selects relay users based on his current resource usage. A bargain game is employed to model the transaction pricing between the bundle carrier and the relay user, which leads to a subgame perfect Nash equilibrium as the agreement of two players to maximize their benefits. The simulation shows that the proposed scheme can efficiently improve both delivery ratio and delay performance in MSNs with an optimal relay service.

4.1 Relay Services in MSNs

In MSNs [1–4], users can create and share their content with each other by using mobile devices equipped with short-range wireless interfaces via peer-to-peer opportunistic links [5–8]. In MSNs, the transmission path between the source and the destination is unstable or even unavailable sometimes [9]. Moreover, since the MSNs have unique features with different types of social ties among mobile nodes, e.g., friends, relatives, etc., the transmission path is even harder to set up than the conventional networks without social ties. Therefore, a store-carry-forward fashion is used to deliver bundles to the destination in MSNs, where this delivery fashion requires mobile nodes to provide relay services.

Although many studies have been carried out to design relay schemes for wireless networks [10–12], most of the relay services assume that the transmission path between the source and the destination always exists and is stable, where these studies cannot be directly applied to MSNs. In addition, the social ties among mobile nodes bring new challenges to the study of bundle delivery in MSNs. Therefore, it is still a new and open problem to design social-aware relay services for bundle delivery in MSNs.

In this chapter, as shown in Fig. 4.1, a novel relay scheme for delivering bundles in MSNs is proposed. First, each node has its own virtual currency and can earn

Z. Su et al., *Modeling and Optimization for Mobile Social Networks*,
DOI 10.1007/978-3-319-47922-4_4

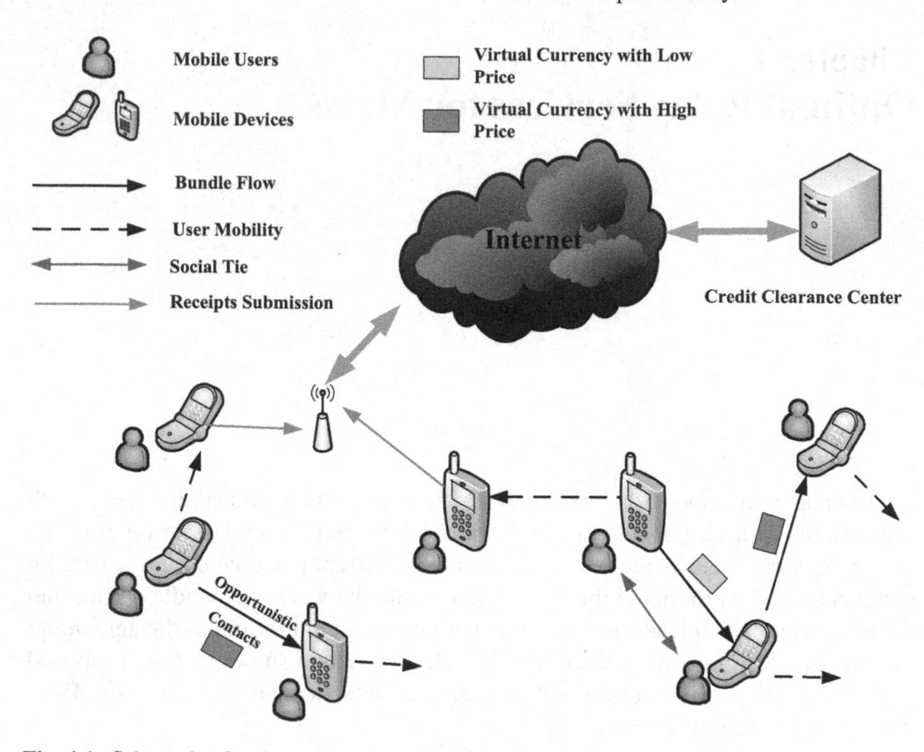

Fig. 4.1 Schematic of an incentive scheme in the MSN

currency as a relay for other nodes. When a node refuses to forward the bundle of others, it will not get paid. Therefore, this node will lose a chance to earn currency to afford the relay service from other nodes in the future, with the result that the proposed scheme can efficiently prevent nodes from being selfish. Next, a bundle carrier selects relay nodes based on its current status of limited resources. Specifically, if the status of the carrier is loose, it will select one of its friends to be a relay with a low agreement price. Otherwise, if the status of the carrier is tense, it will select any node it encounters, i.e., even a nonfriend node, with a high agreement price. Then, a bargain game is employed to model the transaction pricing between the bundle carrier and the relay node, which leads to a subgame perfect Nash equilibrium as the agreement of two players to maximize their benefits. In addition, with simulation experiments, it proves that the proposed scheme is efficient to improve both delivery ratio and delay.

4.2 Related Work

A number of incentive schemes have been proposed to improve the performances of wireless networks. Wang et al. [11] propose a data-sharing scheme by exploiting local historical paths and users' interest information. It can allow nodes to cooperatively deliver information of interest to one another via the chosen paths by utilizing few transmissions in delay tolerant networks. Chen et al. [12] propose a coalitional-gametheory-based incentive scheme to stimulate message forwarding in vehicular ad hoc networks. Wu et al. [13] use the gametheoretic approach to design a novel incentive scheme for stimulating selfish nodes in opportunistic networks, where the end-to-end paths are unstable. Wei et al. [14] introduce a usercentric reputation-based incentive protocol for delay-tolerant networks, where the game-theoretic framework is employed to design costs and the rewarding parameter in bundle forwarding. Mahmoud and Shen [15] propose an incentive system with a payment model in multihop wireless networks, by considering the difference between Web-based applications and cooperation simulation. Ning et al. [16] propose a credit-based incentive scheme to stimulate nodes, where the nodal communication is formulated as a two persons' cooperative game by using the Nash theorem.

Gueguen et al. [17] present an incentive scheduling algorithm where the coverage extension is introduced to motivate and reward nodes' cooperation. Im et al. [18] design an incentive protocol to support content sharing among users in the third-generation/wireless local area network dual-mode networks, which can also encourage the content provider to offer a discounted price for downloading high-quality content. Lee et al. [19] propose a secure incentive protocol to stimulate cooperative diffusion to advertise content over vehicular networks. Tseng et al. [20] design a reed-solomon-code-based incentive scheme to enhance security for vehicular content delivery.

Mobile social applications have attracted much attention. Liang et al. [21] propose a three-step data-forwarding scheme to enable efficient user cooperation and maintain privacy preservation in MSNs by introducing a new concept of social morality as a fundamental social feature of human society. Wang et al. [22] propose a cloud-based multicast scheme with a feedback mechanism in MSNs, which has two phases: pre-cloud and inside-cloud. Niyato et al. [23] present a controlled coalitional game model for interaction between content providers and the network operator to distribute content. Bulut and Szymanski [24] present a friendship-based routing scheme for MSNs, which introduces a novel metric to accurately detect the quality of friendship and make the forwarding decisions. Wu and Wang [25] employ the internal social features of each node for routing, which has two unique processes, including social feature extraction and multipath routing.

Lee and Quek [26] propose a protocol for reliable device-to-device communications, where both the spatial user distribution and the communication distance distribution are considered. Zhang and Cao [27] introduce transient connected-componentaware data-forwarding strategies in MSNs to increase opportunistic contact to enhance the performance of data forwarding. Hu et al. [28] present a distributed multiage cooperative social protocol to disseminate content, where a content owner can multicast content to his social friends. Lin et al. [29] introduce a data-forwarding scheduling model and a back-induction algorithm for prompting nodes to forward messages to appropriate relay nodes.

Different from the aforementioned works, this chapter proposes a novel incentive-driven bundle delivery based on relay selection in MSNs. This work aims to stimulate selfish nodes to participate in data forwarding to improve system performance, including delivery ratio and delivery delay.

4.3 System Model

This section presents the network model, the node model, and the bundle model. The design goal is also illustrated.

4.3.1 Network Model

In the MSN, the routing is carried out in an opportunistic way as the end-to-end connections can not always exist. In the MSN, a source node *Src* sends bundles to a destination node *Dst* depending on relays of intermediate nodes $\{N_1, N_2, \ldots, N_n\}$.

To enable the nodes to pay, there is a Credit Clearance Center (CCC), which is employed to manage the virtual currency for each node [15, 30]. Therefore, before joining the system, every node can register itself to the CCC and obtain its account. Each node should hold a digitally signed receipt for each transaction of relay service and submit the receipt to the CCC. The CCC is a server connected to the Internet, so that the node in the MSN can access the CCC when it connects to the Internet. When the destination receives a bundle and submits ACK to the CCC, the node can get paid after the CCC verifies the receipt. Virtual currency can be used in bundle forwarding to pay for bundle relay service provided by other nodes. If a node does not participate in bundle delivery, it will not get the virtual currency. This means that it will not be able to afford the services from other nodes for its own bundles in the future.

According to the number of message copies in routing, the routing mechanisms can be divided into single-copy and multiple-copy routing. Single-copy routing means that there is only one node having the message copy in the network anytime. Multiple-copy routing is that the message is duplicated to generate multiple copies, and each copy makes routing decisions independently to reduce transmission delay. However, multiple copy routing often consumes and occupies a large amount of network resources. Therefore, the single-copy mechanism is adopted to study the incentive scheme in this chapter.

In this chapter, we focus on the cooperation problem and the incentive scheme to stimulate selfish nodes to participate in data forwarding. Based on related work [15, 31, 32], the mechanism of CCC has already been used by a lot of works, where the overheads for connection and access to the CCC could be controlled and reduced. The reason is that mobile nodes connect to the CCC intermittently and only transfer control messages (receipt and registration) to reduce both the load and overhead, which does not need much power of each node. In addition, in this chapter, as the single-copy mechanism is adopted, it does not need to consume too much energy from the network. There is only one node that has a message, and it needs few connections.

4.3.2 Node Model

The nodes in MSNs are electronic devices that have limited resources, such as buffer, energy, etc. Mobile nodes would exhibit selfishness to save their own resources. The node model is summarized as follows.

- There are two categories of nodes in networks: cooperative nodes and selfish nodes. Since nodes get paid only when the destination receives the bundle, selfish nodes cannot get benefit if they drop bundles.
- Without incentive strategies, if a node is selfish, it accepts no bundles from other nodes unless it is the destination of the bundles.
- Selfish nodes are of limited rational. That is, with the incentive strategies, these nodes pursue for the maximum benefit if nodes have sufficient resources. Meanwhile, nodes only consider whether it is beneficial to accept bundles at this moment and do not consider whether they may break the pale when they want to buy other nodes relay services.

4.3.3 Bundle Model

When a source node *Src* sends a message *M* to a destination *Dst*, *Src* first sets the message head with necessary information and then generates a bundle, as shown in Fig. 4.2.

Specifically, the bundle is comprised of six components: its bundle sequence number ID, source *Src*, destination *Dst*, creating time stamp TS, time-to-live TTL, and message *M*.

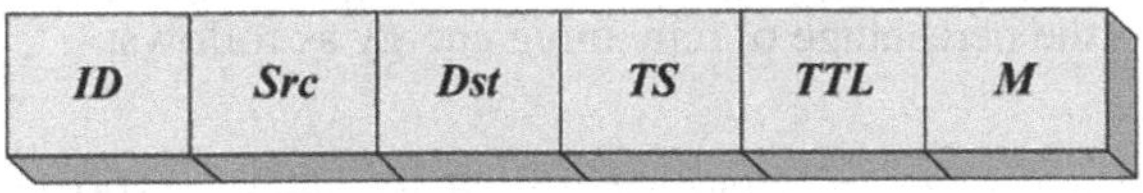

ID	*Src*	*Dst*	*TS*	*TTL*	*M*

Fig. 4.2 Formation of the bundle

4.3.4 Design Goals

Our design goals have two desirable objectives as follows: On one hand, our scheme should be effective in stimulating selfish nodes to participate in bundle delivery in MSNs. On the other hand, it should be efficient without introducing too much extra transmission delay.

4.4 Relay Service Scheme

This section proposes an incentive-driven bundle delivery scheme based on the relay selection. It aims to provide efficient message dissemination in MSNs when selfish nodes exist. First, it introduces an overview of the node status and then provides the detailed node selection strategy. Next, the interaction between the bundle carrier and the relay node is formulated by employing a bargain game. Finally, the detailed bundle-forwarding process between both sides is introduced.

4.4.1 Node Status

By considering the factors that can affect the will of a node to participate in bundle delivery in MSNs, a metric of the node status is elaborated, including node buffer, node energy, and the TTL of the bundle.

- *Buffer* Each node has its limited buffer, and the free space of the buffer gradually decreases with storing more and more data. For simplicity, symbol Bu_i is defined as the percentage of remaining buffer to represent the status of the node on the buffer by

$$Bu_i = \frac{Bu_{re_i}}{Bu_{\max_i}} \times 100\,\%, \tag{4.1}$$

  where Bu_{re_i} is the remaining buffer of node i at present, and $Bu_{\max_i}$ is the maximum buffer of node i.
- *Energy* Similar to the buffer, the energy of each node is also limited. Let E_i denote the percentage of remaining energy as follows:

$$E_i = \frac{E_{re_i}}{E_{\max_i}} \times 100\,\%, \tag{4.2}$$

  where E_{re_i} is the remaining energy of node i at present, and $E_{\max_i}$ is the maximum energy of node i.
- *TTL* The TTL of a bundle has a significant impact on bundle delivery. If the TTL of a certain bundle is going to expire, each node should forward the bundle as soon

as possible. Otherwise, relay nodes will not get paid. The node status on the TTL of the bundle ID_m is defined as

$$T^i_{\mathrm{ID}_m} = \frac{\mathrm{TTL} - (T_c - \mathrm{TS})}{\mathrm{TTL}} \times 100\,\% = \frac{\mathrm{TTL}_{re_i}}{\mathrm{TTL}} \times 100\,\%. \tag{4.3}$$

where $T^i_{\mathrm{ID}_m}$ is the percentage of the remaining TTL of bundle ID_m carried by node i at present. TTL_{re_i} denotes the remaining TTL of the bundle. TS means the creating time stamp of the bundle, and T_c is the current time.

Obviously, at any moment, the given three factors have different impacts on the bundle forwarding of each node. Here, a status metric is introduced by

$$\mathrm{SM}^i_{\mathrm{ID}_m} = \alpha \cdot \log_2(1 + Bu_i) + \beta \cdot \log_2(1 + E_i) + \gamma \cdot \log_2(1 + T^i_{\mathrm{ID}_m}). \tag{4.4}$$

where $\mathrm{SM}^i_{\mathrm{ID}_m}$ is the status metric of node i on bundle ID_m. α, β, and γ are the weight parameters to adjust the importance of buffer, energy, and TTL, respectively, and $\alpha + \beta + \gamma = 1$.

4.4.2 Node Selection

For a bundle carrier, it may always wish that there is a low price to buy a relay node's service. As there are many social ties among nodes, a node is usually willing to help its friends even if it can only get a low benefit. Therefore, bundle owners always wait for friend nodes to forward the bundle. However, when the remaining buffer or energy of a carrier or the remaining TTL of the bundle is little, the gain will be outweighed by the loss for the carrier if it still keeps waiting for its friend nodes. Therefore, bundle owners may select their relay nodes according to their current statuses.

A selection threshold σ_i is defined to determine whether it is appropriate for node i to wait for friends. When node i encounters a relay candidate, it will check its current status. If $\mathrm{SM}^i_{\mathrm{ID}_m} \geq \sigma_i$, which denotes the status of node i is loose, it tends to purchase relay service from its friend nodes, where the agreement price is usually low. If $\mathrm{SM}^i_{\mathrm{ID}_m} < \sigma_i$, which denotes the status of node i is tense, it will forward the bundle to a relay candidate, regardless of whether the candidate is a friend or not.

In fact, when the status is loose, it is unreasonable that a carrier waits for its friend nodes for a long time, as it has few friends or it takes a long time to encounter a friend. Therefore, $L^i_{\mathrm{ID}_m}$ denotes how many candidates node i encounters at most for bundle ID_m when the status is loose, i.e.,

$$L^i_{\mathrm{ID}_m} = \lfloor k_i \mathrm{SM}^i_{\mathrm{ID}_m} \rfloor, \quad \text{for } \mathrm{SM}^i_{\mathrm{ID}_m} > \sigma_i. \tag{4.5}$$

Here, $\lfloor \bullet \rfloor$ represents the floor function, and k_i reflects the change of status, including the attenuation of energy and buffer.

4.4.3 Bargain Game

When a bundle carrier encounters a relay node, the carrier may want to buy the relay node's service, where the bundle carrier is seen as a buyer, and the relay node is modeled as a seller. Both the buyer and the seller usually want to pursue their largest benefits. That is, the buyer hopes that the price of the relay service is low, whereas the seller wants the price to be as high as possible. Therefore, a bargaining game is employed to formulate a pricing model, in the presence of the conflict of interest between buyers and sellers.

These two players of the bargain game are defined by $N = \{B, S\}$, to present the nodes that are the buyer and the seller of a relay service, respectively. The reserve price of bundle ID_m of a buyer B is denoted by $\mathrm{RP}^B_{\mathrm{ID}_m}$ to represent the bearable maximum buying price of a relay service for bundle ID_m. Then, it can be obtained by

$$\mathrm{RP}^B_{\mathrm{ID}_m} = \rho_B \times si_{\mathrm{ID}_m} \times \frac{\mathrm{VC}_B}{\mathrm{SM}^B_{\mathrm{ID}_m}}. \tag{4.6}$$

where ρ_B is the reserve factor of the buyer, and si_{ID_m} is the size of bundle ID_m. VC_B denotes the virtual currency that the buyer currently has. From the above, it can be known that the reserve price of the buyer is higher when the bundle has a larger size or the buyer has more virtual currency. Moreover, the tenser the status of buyer B is, the higher the reserve price becomes.

Similarly, the reserve price of seller S can be denoted by $\mathrm{RP}^S_{\mathrm{ID}_m}$ to represent the bearable minimum selling price of its relay service for bundle ID_m. It can be obtained by

$$\mathrm{RP}^S_{\mathrm{ID}_m} = \rho_S \times si_{\mathrm{ID}_m} \times \frac{\mathrm{VC}_S}{\varepsilon Bu_S + \omega E_S} \times \frac{1}{e^{\eta_{BS}}}, \tag{4.7}$$

where e is Euler's number, and ρ_S is the reserve factor of the seller. Here, VC_S denotes the virtual currency that the seller has. Bu_S means the percentage of the remaining buffer of the seller, and E_S denotes the percentage of the remaining energy of the seller. ε and ω are the weight parameters satisfying $\varepsilon + \omega = 1$. Furthermore, η_{BS} is the friendship factor between buyer B and seller S, which can be shown by

$$\eta_{BS} = \begin{cases} 1, & \textit{if buyer and seller are friends} \\ 0, & \textit{otherwise} \end{cases}. \tag{4.8}$$

Based on (4.7), if the resource of the seller is not enough, it will have low will to relay the bundle (the price is high). If the seller has not enough virtual currency, it will have high will to relay the bundle (the price is low). In addition, a friendship factor is introduced to consider the negotiation between two friends or two strangers. That is, if the buyer and the seller are friends, the present price will be low; otherwise, the present price should be high.

The classical Rubinstein-Stahl bargain game is introduced as a solution to model the interaction between two players who make a bargain to divide a "cake" of size π [33]. They negotiate with each other by proposing offers alternately. In this chapter, the bargain game is employed for modeling the division of the difference value C for the reserve prices of buyer B and seller S, where the difference value C is the "cake". Thus, we have

$$C = \mathrm{RP}^{B}_{\mathrm{ID}_m} - \mathrm{RP}^{S}_{\mathrm{ID}_m}. \tag{4.9}$$

As the players in MSNs may be greedy and selfish, they try to get as much proportion of the "cake" as possible in the bargain game. Their utility functions are denoted as

$$u_S(x_S) = x_S C - R_s, \tag{4.10}$$

$$u_B(x_B) = x_B C - T_B, \tag{4.11}$$

where $u_S(\cdot)$ and $u_B(\cdot)$ are the utility functions of the seller and the buyer. T_B and R_S denote the costs associated with the transmission and reception of the bundle. x_S and x_B mean the proportion of the "cake" divided for the seller and the buyer, respectively. We have

$$X = \left\{(x_S, x_B) \in R^2 : \ x_S + x_B = 1, \ x_S \geq 0, x_B \geq 0\right\}. \tag{4.12}$$

Here, the pair (x_S, x_B) is the offered division by the seller or the buyer.

The bargain procedure between a buyer and a seller is as follows. In round 1, the seller first makes an offer $x_1 = (x_{S1}, x_{B1})$, where $x_1 \in X$. x_{S1} and x_{B1} represent the proportion of the "cake" that the seller and the buyer want, respectively. According to this offer, the buyer can either accept or reject the offer. If the buyer accepts, the agreement is reached, and the bargain game is over. Otherwise, the bargain game comes to round 2, and the buyer then turns to make a new offer $x_2 = (x_{S2}, x_{B2})$, where $x_2 \in X$. x_{S2} and x_{B2} are the proportion of the cake that the seller and the buyer are interested in, respectively. Then, the seller must either accept or reject the new offer that the buyer provides. Similarly, if the seller accepts, the game is over. Otherwise, the game comes to the next round. Therefore, this bargain game is an infinitely repeated game.

Obviously, it takes some cost and time to carry out each round of the negotiation in the bargain game. Therefore, there should be a final agreement accepted by both sides as soon as possible in the negotiation. In other words, each player in the game has its own patience, which is also called the discount factor. This discount factor can depict the utilities of both the buyer and the seller, which decreases over time. We denote δ_S and δ_B as the patience factors of the seller and the buyer, respectively. Therefore, if the patience factors are considered in the game, the utility functions should be as follows:

$$u_S^r(x_S) = \delta_S^{r-1}(x_S C - R_S), \tag{4.13}$$

$$u_B^r(x_B) = \delta_B^{r-1}(x_B C - T_B), \tag{4.14}$$

where $u_S^r(\cdot)$ and $u_B^r(\cdot)$ are the utility functions of the seller and the buyer in round r.

In the bargain game, the patience factor can affect the results of negotiation for both sides. The following presents how to determine the patience factor for the seller and the buyer, respectively. If its status is tenser, buyer B will have lower patience. Moreover, the patience factor is varied from zero to one. Therefore, the conditions for patience functions of buyer B can be defined as follows:

$$\frac{d\delta_B(\mathrm{VC}_S \cdot \mathrm{SM}_{\mathrm{ID}_m}^B)}{d(\mathrm{VC}_S \cdot \mathrm{SM}_{\mathrm{ID}_m}^B)} > 0, \quad \delta_B(0) = 0, \ \delta_B(\infty) = 1, \tag{4.15}$$

where $\mathrm{SM}_{\mathrm{ID}_m}^B$ is the status metric of buyer B on bundle ID_m. The function for the patience of buyer B [34] is defined by

$$\delta_B(t) = \frac{e^{vt} - e^{-vt}}{e^{vt} + e^{-vt}}, \tag{4.16}$$

where v is the patience coefficient of the buyer.

For seller S, due to the limited buffer and energy, it may selfishly behave to pursue maximum benefit. Therefore, if the status of the seller is more loose, it will have less patience for the relay service of a bundle, because it wants to relay more bundles from different nodes to earn more money. If the status of the seller is tenser, it will have higher patience, because it wants to obtain money from the buyer as much as possible. Moreover, the patience factor is from zero to one. Therefore, the conditions for the patience function of seller S are defined as follows:

$$\frac{d\delta_S(\mathrm{VC}_S \cdot (\varepsilon Bu_S + \omega E_S))}{d(\mathrm{VC}_S \cdot (\varepsilon Bu_S + \omega E_S))} < 0, \quad \delta_S(0) = 1, \ \delta_S(\infty) = 0. \tag{4.17}$$

Similarly, the following can be used to express the patience function [34] for seller S:

$$\delta_S(t) = 1 - \frac{e^{\mu t} - e^{-\mu t}}{e^{\mu t} + e^{-\mu t}}, \tag{4.18}$$

where μ is the patience coefficient of the seller.

According to the given description, to avoid the loss on negotiation, each player prefers reaching agreements as soon as possible. Then, we have the following theorems.

Theorem 1 *In the proposed bargain game, there exists a unique subgame perfect Nash equilibrium. According to Nash equilibrium, the bargain ends in one round with the following agreement:*

$$x_S^* = \frac{C - \delta_B C - \delta_B \delta_S R_S + \delta_B R_S + \delta_B T_B - T_B}{C - \delta_B \delta_S C}. \tag{4.19}$$

$$x_B^* = \frac{\delta_B C - \delta_B \delta_S C + \delta_B \delta_S R_S - \delta_B R_S - \delta_B T_B + T_B}{C - \delta_B \delta_S C}. \tag{4.20}$$

Proof A subgame Nash perfect equilibrium is constructed by backward induction.

Since this bargain game is infinitely repeated, it is hard to find the point from which the back induction is started. Therefore, there should have been a subgame Nash perfect equilibrium in the game, where the seller makes an offer x^* and the buyer accepts this offer in round 1. Similarly, if the game begins from round 3, the seller will give the offer x^*, and the buyer will accept this offer. Therefore, this infinitely repeated game can be seen as a bargain game with three rounds to analyze and obtain the equilibrium. In the following analysis, the offer x denotes the proportion of the difference value C that seller S can get, where $x_S = x$, $x_B = 1 - x$.

Starting from round 3, the last mover should be buyer B. It will accept proposal x^* offered by seller S. Next, move to round 2, where buyer B is to give an offer. Seller S accepts the proposal x_2 only when

$$\begin{cases} u_S^2 \geq u_S^3 \\ \delta_S(x_2 C - R_S) \geq \delta_S^2(x^* C - R_S) \\ x_2 \geq \dfrac{\delta_S x^* C - \delta_S R_S + R_S}{C} \end{cases}. \tag{4.21}$$

Then, buyer B makes a proposal x_2^* that maximizes its utility. Thus, we can derive

$$\begin{aligned} x_2^* &= \arg\max_{x_2}(u_B^2) \\ &= \arg\max_{x_2} \begin{cases} \delta_B[(1 - x_2)C - T_B], & if\ x_2 \geq \dfrac{\delta_S x^* C - \delta_S R_S + R_S}{C} \\ u_B^3, & if\ x_2 < \dfrac{\delta_S x^* C - \delta_S R_S + R_S}{C} \end{cases} \\ &= \frac{\delta_S x^* C - \delta_S R_S + R_S}{C} \end{aligned}. \tag{4.22}$$

Next, let us move to round 1 where seller S is to give an offer. Buyer B accepts a proposal only when

$$\begin{cases} u_B^1 \geq u_B^2 \\ (1 - x_1)C - T_B \geq \delta_B[(1 - x_2^*)C - T_B] \\ x_1 \leq \dfrac{C - \delta_B[(1 - x_2^*)C - T_B] - T_B}{C} \\ x_1 \leq \dfrac{C - \delta_B C + \delta_B \delta_S x^* C - \delta_B \delta_S R_S + \delta_B R_S + \delta_B T_B - T_B}{C} = \Delta \end{cases}. \tag{4.23}$$

Then, seller S makes a proposal x_1^* that maximizes its utility. Thus, we can derive

$$\begin{aligned} x_1^* &= \arg\max_{x_1}(u_S^1) \\ &= \arg\max_{x_1} \begin{cases} x_1 C - R_S, & if\ x_1 \le \Delta \\ u_S^2, & if\ x_1 > \Delta \end{cases} \\ &= \frac{C - \delta_B C + \delta_B \delta_S x^* C - \delta_B \delta_S R_S + \delta_B R_S + \delta_B T_B - T_B}{C} \end{aligned} \quad (4.24)$$

As the equilibrium of the infinitely repeated bargain game is equal to the equilibrium of the bargain game with three rounds, it can be obtained by

$$\begin{cases} x_1^* = x^* \\ \dfrac{C - \delta_B C + \delta_B \delta_S x^* C - \delta_B \delta_S R_S + \delta_B R_S + \delta_B T_B - T_B}{C} = x^* \\ x^* = \dfrac{C - \delta_B C - \delta_B \delta_S R_S + \delta_B R_S + \delta_B T_B - T_B}{C - \delta_B \delta_S C} \end{cases} . \quad (4.25)$$

Therefore, the subgame Nash perfect equilibrium becomes

$$x_S^* = \frac{C - \delta_B C - \delta_B \delta_S R_S + \delta_B R_S + \delta_B T_B - T_B}{C - \delta_B \delta_S C}. \quad (4.26)$$

$$x_B^* = \frac{\delta_B C - \delta_B \delta_S C + \delta_B \delta_S R_S - \delta_B R_S - \delta_B T_B + T_B}{C - \delta_B \delta_S C}. \quad (4.27)$$

This completes our proof.

Theorem 2 *The transaction price TP of a relay service for bundle* ID_m *is*

$$\mathrm{TP} = \mathrm{RP}^B_{\mathrm{ID}_m} - x_B^* C. \quad (4.28)$$

Meanwhile, if the buyer and the seller are two friends, the transaction price will be lower than that if they are not friends.

Proof First, we analyze the scenario that the buyer and the seller are friends. With a transaction price TP_f, it can be obtained by

$$\mathrm{RP}^B_{\mathrm{ID}_m} - \mathrm{TP}_f - T_B = x_B^* C_f - T_B. \quad (4.29)$$

$$\mathrm{TP}_f = \mathrm{RP}^B_{\mathrm{ID}_m} - x_B^* C_f. \quad (4.30)$$

Here, C_f is the difference value of buyer B and seller S who are friends.

According to (4.6)–(4.8), C_f can be calculated as follows:

$$\begin{aligned} C_f &= \mathrm{RP}^B_{\mathrm{ID}_m} - \mathrm{RP}^S_{\mathrm{ID}_m} \\ &= \rho_B \times si_{\mathrm{ID}_M} \times \frac{\mathrm{VC}_B}{\mathrm{SM}^B_{\mathrm{ID}_m}} - \rho_S \times si_{\mathrm{ID}_M} \times \frac{\mathrm{VC}_S}{\varepsilon B u_S + \omega E_S} \times \frac{1}{e}. \end{aligned} \tag{4.31}$$

Therefore, if the player and the seller are friends, the transaction price TP_f will be

$$\begin{aligned} \mathrm{TP}_f &= \mathrm{RP}^B_{\mathrm{ID}_m} - x^*_B \left(\rho_B \times si_{\mathrm{ID}_M} \times \frac{\mathrm{VC}_B}{\mathrm{SM}^B_{\mathrm{ID}_m}} - \rho_S \times si_{\mathrm{ID}_M} \times \frac{\mathrm{VC}_S}{\varepsilon B u_S + \omega E_S} \times \frac{1}{e} \right). \\ &= \mathrm{RP}^B_{\mathrm{ID}_m} - \frac{\delta_B C_f - \delta_B \delta_S C_f + \delta_B \delta_S R_S - \delta_B R_S - \delta_B T_B + T_B}{1 - \delta_B \delta_S} \end{aligned} \tag{4.32}$$

Then, we analyze the scenario that the buyer and the seller are not friends. With a transaction price TP_{nf}, we can derive

$$\mathrm{RP}^B_{\mathrm{ID}_m} - \mathrm{TP}_{nf} - T_B = x^*_B C_{nf} - T_B. \tag{4.33}$$

$$\therefore \mathrm{TP}_{nf} = \mathrm{RP}^B_{\mathrm{ID}_m} - x^*_B C_{nf}. \tag{4.34}$$

Here, C_{nf} is the difference value of buyer B and seller S, who are not friends.

According to (4.6)–(4.8), C_{nf} can be calculated as follows:

$$\begin{aligned} C_{nf} &= \mathrm{RP}^B_{\mathrm{ID}_m} - \mathrm{RP}^S_{\mathrm{ID}_m} \\ &= \rho_B \times si_{\mathrm{ID}_M} \times \frac{\mathrm{VC}_B}{\mathrm{SM}^B_{\mathrm{ID}_m}} - \rho_S \times si_{\mathrm{ID}_M} \times \frac{\mathrm{VC}_S}{\varepsilon B u_S + \omega E_S}. \end{aligned} \tag{4.35}$$

Therefore, if the player and the seller are not friends, the transaction price TP_{nf} becomes

$$\begin{aligned} \mathrm{TP}_{nf} &= \mathrm{RP}^B_{\mathrm{ID}_m} - x^*_B \left(\rho_B \times si_{\mathrm{ID}_M} \times \frac{\mathrm{VC}_B}{\mathrm{SM}^B_{\mathrm{ID}_m}} - \rho_S \times si_{\mathrm{ID}_M} \times \frac{\mathrm{VC}_S}{\varepsilon B u_S + \omega E_S} \right). \\ &= \mathrm{RP}^B_{\mathrm{ID}_m} - \frac{\delta_B C_{nf} - \delta_B \delta_S C_{nf} + \delta_B \delta_S R_S - \delta_B R_S - \delta_B T_B + T_B}{1 - \delta_B \delta_S} \end{aligned} \tag{4.36}$$

Then, the comparison between TP_f and TP_{nf} is carried out by

$$
\begin{aligned}
\mathrm{TP}_f - \mathrm{TP}_{nf} &= \mathrm{RP}^B_{\mathrm{ID}_m} - \frac{\delta_B C_f - \delta_B \delta_S C_f + \delta_B \delta_S R_S - \delta_B R_S - \delta_B T_B + T_B}{1 - \delta_B \delta_S} \\
&\quad - \mathrm{RP}^B_{\mathrm{ID}_m} + \frac{\delta_B C_{nf} - \delta_B \delta_S C_{nf} + \delta_B \delta_S R_S - \delta_B R_S - \delta_B T_B + T_B}{1 - \delta_B \delta_S}. \\
&= \frac{\delta_B (1 - \delta_S)(C_{nf} - C_f)}{1 - \delta_B \delta_S} \\
&= \frac{\delta_B (1 - \delta_S)}{1 - \delta_B \delta_S} \times \rho_S \times si_{\mathrm{ID}_M} \times \frac{\mathrm{VC}_S}{\varepsilon B u_S + \omega E_S} (\frac{1}{e} - 1) < 0 \qquad (4.37)
\end{aligned}
$$

$$
\mathrm{TP}_f < \mathrm{TP}_{nf}. \qquad (4.38)
$$

Therefore, if the buyer and the seller are friends, the transaction price is lower than that if they are not. This completes our proof.

4.4.4 *Bundle Delivery Framework*

(1) When the carrier of bundle ID_m encounters a node, whose probability of encountering the destination is higher than that of the carrier, the carrier first checks its current status by (4.4). If the status is loose, the carrier will check whether the encountering node is its friend node or not. If this node is a friend node, the carrier will send the requesting information, including the bundle ID, the size of the bundle, the reserve price [calculated by (4.6)], and the patience factor [calculated by (4.16)]. If this node is not a friend node, the carrier will wait for other friend nodes until the number of encountering nodes is equal to the maximum number decided by (4.5). If the status of the carrier is tense, the carrier will immediately send the requesting information.
(2) When the relay node receives the requesting information, if the relay node is not the destination, it will calculate its reserve price by (4.7). If the reserve price of the relay node is lower than that of the bundle carrier, the relay node calculates the transaction price by (4.7), (4.8), (4.18), (4.20), and (4.21). Then, the relay node will send acknowledgment information back to the bundle carrier, including the bundle ID, the transaction price, the reserve price, and the patience factor of the relay node.
(3) The bundle carrier will check whether the obtained utility is positive or not. If $u^1_B > 0$, the bundle carrier will accept the price and send bundle ID_m to the relay node. Otherwise, it will wait for another node.
(4) After the relay node receives the bundle, both sides of the transaction sign a digital receipt, which includes the bundle ID and price, and each side holds a copy of this receipt. The receipts will be submitted to the CCC when they connect to the Internet. Then, the bundle carrier deletes the bundle ID_m in its buffer.
(5) When the destination receives the bundle, the destination submits an ACK to the CCC including the bundle ID when it connects to the Internet. Then, the CCC pays the corresponding virtual currency to each relay node based on the digital receipts.

4.5 Performance Evaluations

In this section, the performance of the proposed incentive scheme is evaluated. We introduce the simulation setup and show the performance comparison with discussions.

4.5.1 Simulation Setup

In the simulation, there are 20 nodes with a transmission radius of 50 m, which are uniformly deployed in an area of 1 km × 1 km. Each node moves at a speed uniformly spread in [0.5, 2.5] m/s with the random direction model. Therefore, according to [35], the average contact rate of two nodes is 0.37 contacts per hour to determine the contact time of each pair of nodes for the simulation in MATLAB.

The source node generates bundles with a uniform time interval of 10 min, and the size of each bundle is 2 MB. We set the buffer size of each node to 30 MB. The TTL of the bundle is 6 h. The destination of the bundle is randomly selected from the other nodes except source nodes. The energy of each node decreases according to $E_{re} = e^{-\lambda t} E_{\max}$, where $\lambda = 0.2$. Moreover, the weight parameters are $\alpha = 0.3$, $\beta = 0.3$, and $\gamma = 0.4$. Table 4.1 lists the parameter values in this simulation.

The social ties among mobile nodes are generated by using the BA model [36], which can generate a scale-free social network model. Each simulation runs for 12 h and is repeated ten times. Every node has an initial virtual currency of 100 and pays the corresponding currency for each delivered bundle.

Table 4.1 Parameters

Parameter	Value
N: the number of users in the network	20
$Bu_{\max}$: the maximum buffer of each node	30 MB
$E_{\max}$: the maximum energy of each node	2000 w
TTL: the time-to-live value of a bundle	6 h
$\{\alpha, \beta, \gamma\}$: the weight parameters in (4.4)	{0.3,0.3,0.4}
σ: the section threshold of each node	0.7
k: the change of status	3
$\{\varepsilon, \omega\}$: the weight parameters in (4.7)	{0.5,0.5}
ν: the patience coefficient of the buyer	0.6
μ: the patience coefficient of the seller	0.6
$\{T_B, R_S\}$: the costs associated with the transmission and recepition of bundle, respectively	{0, 0}

The following metrics are used to compare different delivery schemes:

- *delivery ratio* the proportion of the bundles that have been delivered to the bundles being created;
- *delivery delay* the average delivery time that is used to deliver bundles from the source to the destination.

4.5.2 Performance Comparison

The incentive scheme is compared with three conventional delivery schemes as follows.

Epidemic [37] In this scheme, bundles are flooded when a bundle carrier encounters other nodes that do not possess a copy of the bundle.

Direct Deliver [38] In this scheme, the source holds the bundle until it comes in contact with the destination.

PRoPHET+ [39] In this scheme, the carrier forwards the bundle to other nodes with the weighted function determined by the nodes buffer size, power, and predictability.

Figure 4.3 shows the delivery ratio by comparing the proposed scheme with other three schemes. In Fig. 4.3, it is shown that the proposed scheme outperforms other existing delivery schemes when the number of selfish nodes in MSNs changes. In the epidemic scheme, since there is no incentive strategy, selfish nodes refuse to relay bundles. Therefore, it causes that many bundles are dropped when bundles are expired or the buffer of the node is overflowing. In PRoPHET+, the bundles cannot be delivered to the destination with a high delivery ratio due to the selfish nodes. For the direct-delivery scheme, it shows the worst performance since it only delivers the bundle when arriving at the destination.

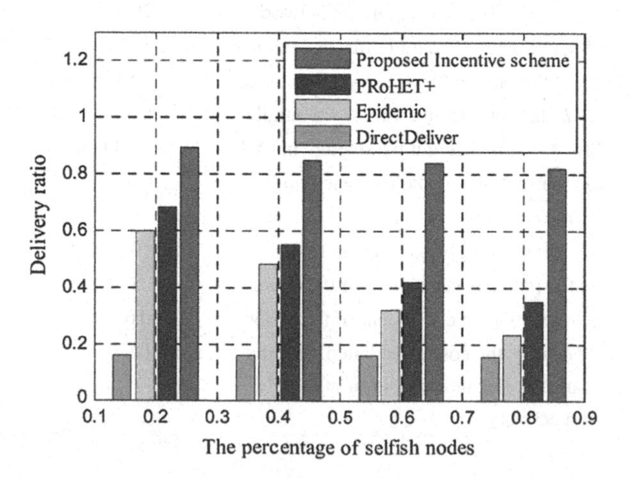

Fig. 4.3 Delivery ratio versus percentage of selfish nodes

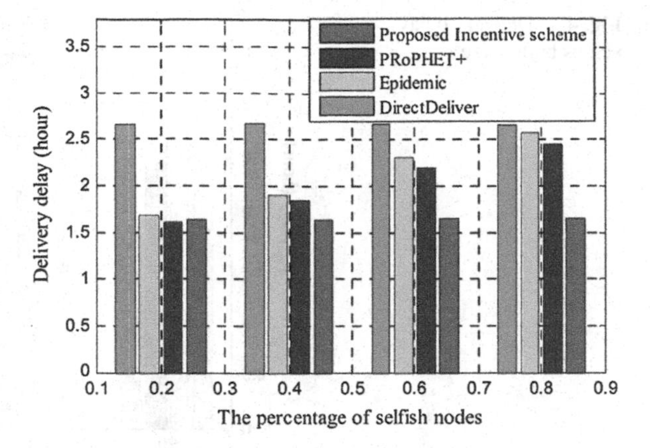

Fig. 4.4 Delivery delay versus percentage of selfish nodes

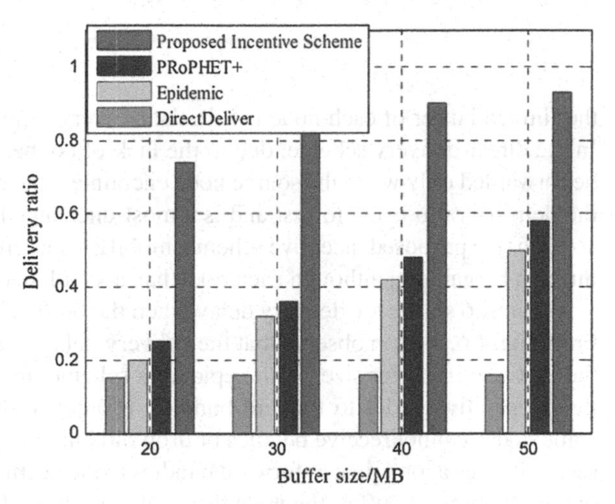

Fig. 4.5 Delivery ratio versus buffer size

Figure 4.4 shows the comparison of delivery delay. From this figure, we can infer that the proposed incentive scheme has the lowest delay as almost all nodes participate in bundle delivery. In epidemic, because some nodes are still selfish, bundle owners are required to wait for cooperative nodes. In PRoPHET+, the carriers have to wait for the cooperation nodes for bundle forwarding, thus resulting in longer delay. For direct-delivery, it takes a long time to encounter the destination, thus resulting in the largest delay.

Next, we test the delivery ratio and delay with different buffer sizes, where the buffer size is changed from 20 to 50 MB, and the percentage of selfish nodes is fixed to 0.6. In Fig. 4.5, although all of the delivery ratios of the three schemes increase when the buffer size of each node is increased, the proposed scheme has the maximum delivery ratio. In the epidemic scheme, bundles are stored in the buffer of a node for a long time, thus resulting in the fact that many bundles cannot be forwarded when the buffer is overflowing. In PRoPHET+, as many bundles may be dropped when

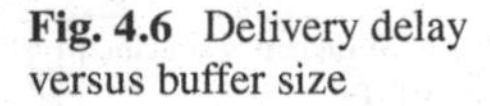
Fig. 4.6 Delivery delay versus buffer size

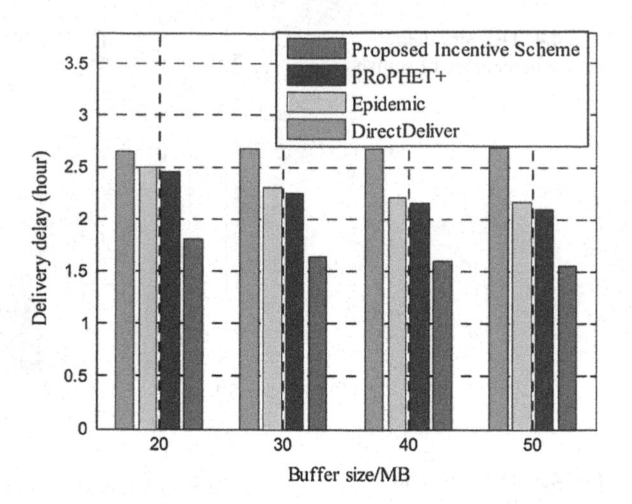

the limited buffer of each node is full, the delivery ratio cannot be improved much. In the direct-delivery scheme, due to the lack of cooperation of nodes, bundles can be forwarded only when the source node encounters the destination node. Therefore, the delivery ratio is the lowest and is almost unchanged with the increase in buffer size. For the proposed incentive scheme, mobile nodes are willing to forward bundles through bargaining, although each node has a small-sized buffer.

Figure 4.6 shows the delivery delay when the buffer size of each node is changed. From Fig. 4.6, we can observe that the delivery delay of each scheme decreases with the increase in buffer size. In the epidemic scheme, mobile nodes have to wait for the cooperative nodes to forward bundles. Moreover, due to the small buffer size, some nodes cannot receive bundles or drop old bundles when the buffer is full. As a result, it takes a long time to forward bundles to the destination node. In PRoPHET+, due to the limited buffer, the node drops old bundles when the buffer is full, where the delivery delay cannot be efficiently reduced. In the direct-delivery scheme, the forwarding of a bundle depends on the encounter between the source node and the destination node, which takes a long time. Compared with the other three schemes, the delivery delay of the proposed scheme is the minimum. In the proposed scheme, although the buffer size is limited, the mobile nodes are willing to forward bundles with an appropriate transaction price, resulting in the fact that the bundle can reach the destination quickly.

The experiments have shown that the proposed incentive scheme can achieve the largest delivery ratio and the lowest delivery delay, compared with other existing protocols when the number of selfish nodes is changed. In addition, when the buffer of each node is changed, our incentive scheme can also obtain better performances in term of both the delivery ratio and delivery delay than others. According to these results, our proposed scheme can outperform other existing algorithms.

4.6 Summary

This chapter has presented a game-theoretic scheme for relay selection to stimulate nodes to participate in bundle delivery in MSNs. In the proposed scheme, the bundle carrier can select a friend node or a stranger to be a relay node based on its status and then pay for the corresponding virtual currency for relay service. In addition, the transaction pricing can be decided by a bargain game, where a subgame Nash perfect equilibrium is used to calculate the agreement price. Extensive simulations show that the proposed scheme can outperform other existing schemes with a higher delivery ratio and lower delivery delay. Future work will entail the security issues of payment if there are some fraudulent nodes in MSNs.

References

1. N. Kayastha, D. Niyato, P. Wang, E. Hossain, Applications, architectures, and protocol design issues for mobile social networks: a survey. Proc. IEEE **99**(12), 2125–2129 (2011)
2. Z. Su, Q. Xu, H. Zhu, Y. Wang, A novel design for content delivery over software defined mobile social networks. IEEE Netw. **29**(4), 62–67 (2015)
3. K. Zhu, W. Li, X. Fu, Smart: a social- and mobile-aware routing strategy for disruption-tolerant networks. IEEE Trans. Veh. Technol. **63**(7), 3423–3434 (2014)
4. Q. Xu, Z. Su, K. Zhang, P. Ren, Epidemic information dissemination in mobile social networks with opportunistic links. IEEE Trans. Emerg. Top. Comput. **3**(3), 399C–409 (2015)
5. Z. Su, Q. Xu, Content distribution over content centric mobile social networks in 5G. IEEE Commun. Mag. **53**(6), 66–72 (2015)
6. Y. Liu, Z. Yang, T. Ning, H. Wu, Efficient quality-of-service (QoS) support in mobile opportunistic networks. IEEE Trans. Veh. Technol. **63**(9), 4574–4584 (2014)
7. Q. Xu, Z. Su, B. Han, D. Fang, Z. Xu, X. Gan, Analytical model with a novel selfishness division of mobile nodes to participate cooperation. Peer-to-Peer Netw. Appl. **9**(4), 1–9 (2015)
8. X. Zhang, Z. Zhang, J. Xing, R. Yu, Exact outage analysis in cognitive two-way relay networks with opportunistic relay selection under primary user's interference. IEEE Trans. Veh. Technol. **64**(6), 2502–2511 (2015)
9. K. Fall, A delay-tolerant network architecture for challenged internets, in *Proceedings of ACM SIGCOMM* (New York, 2003), pp. 27–34
10. J. Hubaux, Security and cooperation in wireless networks, in *Proceedings of ESAS* (Beijing, 2007), pp. 43–49
11. Y. Wang, M. Chuah, Y. Chen, Incentive based data sharing in delay tolerant mobile networks. IEEE Trans. Wirel. Commun. **13**(13), 370–381 (2014)
12. T. Chen, L. Zhu, F. Wu, S. Zhong, Stimulating cooperation in vehicular ad hoc networks: a coalitional game theoretic approach. IEEE Trans. Veh. Technol. **60**(2), 566–579 (2011)
13. F. Wu, T. Chen, S. Zhong, C. Qiao, G. Chen, A game-theoretic approach to stimulate cooperation for probabilistic routing in opportunistic networks. IEEE Trans. Wirel. Commun. **12**(4), 1573–1583 (2013)
14. L. Wei, Z. Cao, H. Zhu, MobiGame: a user-centric reputation based incentive protocol for delay/disruption tolerant networks, in *Proceedings of IEEE GLOBECOM* (Houston, 2011), pp. 1–5
15. M. Mahmoud, X. Shen, PIS: a practical incentive system for multihop wireless networks. IEEE Trans. Veh. Technol. **59**(8), 4012–4025 (2010)
16. T. Ning, Z. Yang, X. Xie, H. Wu, Incentive-aware data dissemination in delay-tolerant mobile networks, in *Proceedings of IEEE SECON* (Salt Lake City, 2011), pp. 539–547

17. C. Gueguen, A. Rachedi, M. Guizani, Incentive scheduler algorithm for cooperation and coverage extension in wireless networks. IEEE Trans. Veh. Technol. **62**(2), 797–808 (2012)
18. H. Im, Y. Lee, S. Bahk, Incentive-driven content distribution in wireless multimedia service networks, in *Proceedings of IEEE GLBECOM* (Miami, 2010), pp. 1–5
19. S. Lee, J. Park, M. Gerla, S. Lu, Secure incentives for commercial ad dissemination in vehicular networks. IEEE Trans. Veh. Technol. **61**(6), 150–159 (2007)
20. F. Tseng, Y. Liu, J. Hwu, R. Chen, A secure reedcsolomon code incentive scheme for commercial ad dissemination over vanets. IEEE Trans. Veh. Technol. **60**(9), 4598–4608 (2011)
21. X. Liang, X. Li, T. Luan, R. Lu, X. Lin, X. Shen, Morality-driven data forwarding with privacy preservation in mobile social networks. IEEE Trans. Veh. Technol. **61**(7), 3209–3222 (2012)
22. Y. Wang, J. Wu, W. Yang, Cloud-based multicasting with feedback in mobile social networks. IEEE Trans. Wirel. Commun. **12**(12), 6043–6053 (2013)
23. D. Niyato, P. Wang, W. Saad, A. Hjorungnes, Controlled coalitional games for cooperative mobile social networks. IEEE Trans. Veh. Technol. **60**(4), 1812–1824 (2011)
24. E. Bulut, B. Szymanski, Friendship based routing in delay tolerant mobile social networks, in *Proceedings of IEEE GLOBECOM* (Miami, 2010), pp. 1–5
25. J. Wu, Y. Wang, Social feature-based multi-path routing in delay tolerant networks, in *Proceedings of IEEE INFOCOM* (Orlando, 2012), pp. 1368–1376
26. J. Lee, T. Quek, Device-to-device communication in wireless mobile social networks, in *Proceedings of IEEE VTC* (Seoul, 2014), pp. 1–5
27. X. Zhang, G. Cao, Efficient data forwarding in mobile social networks with diverse connectivity characteristics, in *Proceedings of IEEE ICDCS* (Madrid, 2014), pp. 31–40
28. J. Hu, L. Yang, L. Hanzo, Distributed cooperative social multicast aided content dissemination in random mobile networks. IEEE Trans. Veh. Technol. **64**(7), 3075–3089 (2015)
29. C. Lin, W. Lin, C. Chou, Social-based content diffusion in pocket switched networks. IEEE Trans. Veh. Technol. **60**(9), 4539–4548 (2011)
30. S. Zhou, F. Wu, On designing collusion-resistant routing schemes for non-cooperative wireless ad hoc networks. IEEE Trans. Netw. **18**(2), 582–595 (2010)
31. F. Wu, T. Chen, S. Zhong, C. Qiao, G. Chen, A bargaining-based approach for incentive-compatible message forwarding in opportunistic networks, in *Proceedings of IEEE ICC* (Ottawa, 2012), pp. 789–793
32. D. Wen, Y. Cai, Z. Li, Y. Fan, An incentive compatible two-hop multi-copy routing protocol in DTNs, in *Proceedings of IEEE MSN* (Dalian, 2013), pp. 140–146
33. A. Rubinstein, M. Osborne, *Bargain and Markets* (Academic, San Diego, 1990)
34. V. Le, Z. Feng, D. Bourse, P. Zhang, A cell based dynamic spectrum management scheme with interference mitigation for cognitive networks, in *Proceedings of IEEE VTC* (Singapore, 2008), pp. 1594–1598
35. R. Groenvelt, *Stochastic Models in Mobile Ad Hoc Networks* (Univ. Nice Press, Nice, 2005)
36. R. Albert, A. Barabsi, Statistical mechanics of complex networks. Rev. Mod. Phys. **74**(1), 47–97 (2002)
37. S. Ali, J. Qadir, A. Baig, Routing protocols in delay tolerant networks—a survey, in *Proceedings of ICET* (Islamabad, 2010), pp. 70–75
38. A. Socievole, F. Rango, C. Coscarella, Routing approaches and performance evaluation in delay tolerant networks, in *Proceedings of WTS* (New York, 2011), pp. 1–6
39. T. Huang, C. Lee, L. Chen, Prophet+: an adaptive prophet-based routing protocol for opportunistic network, in *Proceedings of IEEE AINA* (Perth, 2010), pp. 112–119

Chapter 5
Optimal Cloud Resource Allocation for MSNs

In this chapter, we study the optimization for cloud resource allocation in MSNs to provide mobile social services with high capacity and low latency. Here, the media cloud means the cloud which can provide users with resources such as media processing or storage to process media tasks. For example, some complicated computation or large-sized content storage may need a large amount of resources which take a big burden for users. The media cloud performs these computation or storage at the side of media cloud, where the required resource at the side of users can be reduced. Specifically, the media cloud sells cloud resources to brokers to obtain revenue. The brokers employ the cloud resource to process media tasks for users. Users determine their brokers to connect to obtain cloud service according to the competition with each other. To model the interactions among media cloud, brokers, and users on cloud resource, the resource allocation problem is formulated by a four-stage Stackelberg game. In addition, an iteration algorithm is proposed to obtain the Stackelberg equilibrium.

5.1 Cloud Resource Allocation in MSNs

MSNs allow more users to have interactions with each other and obtain various multimedia content [1–3]. Recent studies [4] show that the number of users keeps increasing and the traffic of mobile data will be nearly tenfold in 2019, compared with that in 2014. Especially, with the popularity of shared data plan in the near future, users may not only obtain but also share more multimedia contents with others who have social relations with them [5–7]. Therefore, providing users with efficient multimedia services becomes more important than before [5, 8, 9].

However, to provide users with satisfied multimedia services, there exist some new problems to be resolved. On one hand, due to the explosive growth of volume of multimedia and the high demand of quality of experience (QoE), providing users with multimedia services [10–12] needs a large amount of resource. However, the local mobile devices in users always have the limited resource, such as capacity, bandwidth,

Z. Su et al., *Modeling and Optimization for Mobile Social Networks*,
DOI 10.1007/978-3-319-47922-4_5

buffer, etc. New consideration is needed to reduce the consumed resource. On the other hand, multimedia content servers are remotely placed from users. It takes time for users to obtain the requested multimedia content, resulting in a further QoE degradation. For example, if a user wants to watch a movie with his mobile device, the content of movie has to be retrieved from a remote multimedia content server through a large number of routing nodes.

To resolve the above issues, media cloud has been advocated with the following reasons [13]. Firstly, media cloud can deploy cloud resource to process multimedia tasks. Some complicated computations or large-sized multimedia content storage requiring extra resource can be performed at the side of media cloud, where the required resource can be reduced for users. Therefore, the media cloud can help users to save their resource. Secondly, a broker [14] can be placed between media cloud and users. As the broker can act as a proxy which is close to users, users can connect media cloud through the broker for obtaining multimedia services. With the high-speed communication links between media cloud and the broker, users can obtain multimedia services faster than contacting the remote multimedia content servers.

As the resource to be allocated among media cloud, brokers and users is limited, resource allocation becomes a very important challenge. However, the conventional resource allocation schemes can not be directly used to allocate resource among these three parties. First, there exist some significant social features in media cloud with users. For example. users within the same community may have the same interest and social activities [15–17], resulting in the similar demand of QoE for multimedia services. Therefore, social features should be considered to determine the resource allocation. Besides, users in the same community can know the information of each other. Thus, the decision of a user on the selection of broker may be influenced by others. As a result, the affection and competition among different parties should also be taken into consideration for resource allocation.

Although some related studies have been carried out to study resource allocation about cloud computing and mobile networks [18, 19], few of them have studied the resource allocation problem based on the social features in media cloud. In addition, most of them mainly focus on behaviors of servers, instead of three parties including media cloud, brokers and users. Therefore, it is still a new and open problem to design resource allocation scheme of media cloud with users.

5.2 Related Work

5.2.1 Multimedia Social Networks with Mobile Users

Recently, there has been an increasing interest in studying models and schemes for multimedia social network with mobile users. Chang et al. [20] present a general architecture of MSNs where the major components are client devices, wireless access network, and server, to increase the social connection and improve the quality of social service. Wu and Yang [21] design a novel routing scheme by considering the internal social features of nodes, including both social feature extraction and multi-path routing. Wang et al. [22] present a cloud-based multicast scheme in MSNs, where the message forwarding strategy is based on a metric iteratively refined from the feedback control mechanism. Lu et al. [23] design a community based distributed set-cover algorithm to identify the users who have the maximum influence on information diffusion in MSNs.

Li et al. [24] propose a novel data forwarding approach with a space-crossing community detection method to improve the data forwarding efficiency in MSNs. Yin et al. [25] propose a photography model to assist mobile users for capturing high quality photos with mobile devices and crowd sourced social media [26]. Wang et al. [27] propose a novel mobile streaming framework with two main parts: adaptive mobile video streaming and efficient social video sharing. Xu et al. [28] develop an analytical model to mimic information dissemination among users. By introducing pre-immunity and immunity elements, the proposed model can show the change of mobile nodes' interests during information dissemination efficiently. Although these works have studied several aspects of multimedia social networks with mobile users, the details on resource allocation problem with media cloud have not been given.

5.2.2 Resource Allocation with Media Cloud

The resource allocation in media cloud have been studied extensively. Alasaad et al. [29] propose an algorithm for resource reservation in media cloud based on the prediction of demand for streaming capacity. It can maximally exploit the discounted rates offered in the tariffs, while ensuring the sufficient resource to be reserved. Hong et al. [30] present a media task QoS based resource allocation algorithm in media cloud, by considering the service satisfaction of multitask. Magedanz et al. [31] evaluate the effects of multiple factors in a large-scale cloud environment, by defining the metric for assessing the performance of cloud brokering systems. Yin et al. [32] study the operations of cloud computing and wireless networks in mobile computing environments by considering not only the spectrum efficiency but also the pricing information in the cloud.

Sardis et al. [33] introduce a novel concept of cloud-based mobile media service delivery in which services run on the localized public clouds. Ren and Schaar [34] develop an online algorithm that cloud operator can dynamically adjust the resource provisioning according to the time-varying wireless channel conditions. Aggarwal et al. [35] introduce a generalized framework to compute the amount of resource to support media services with a generic cost function. Lu et al. [36] propose a service provisioning model to manage the resources in the hybrid cloud where the profit can be maximized. Xu et al. [37] present an incentive scheme for the relay selection to encourage selfish mobile nodes to participate in bundle delivery, where the relay resource can be allocated based on a game theoretical model. Although the above works have made a lot of efforts for the resource allocation, the characteristics of mobile users have not been considered enough. In addition, how to efficiently use cloud brokers to allocate cloud resource has not been mentioned either.

5.3 System Model

5.3.1 System Model

As shown in Fig. 5.1, there are three parties which are media cloud, brokers, and users within the communities, respectively. The media cloud is composed of a large number of servers which can be used to compute, store, and provide media contents and media application. The brokers can be seen as proxies to process the media tasks of users, where the brokers receive the media tasks from users and then buy the corresponding resources to process the tasks. Users with the similar interest can form

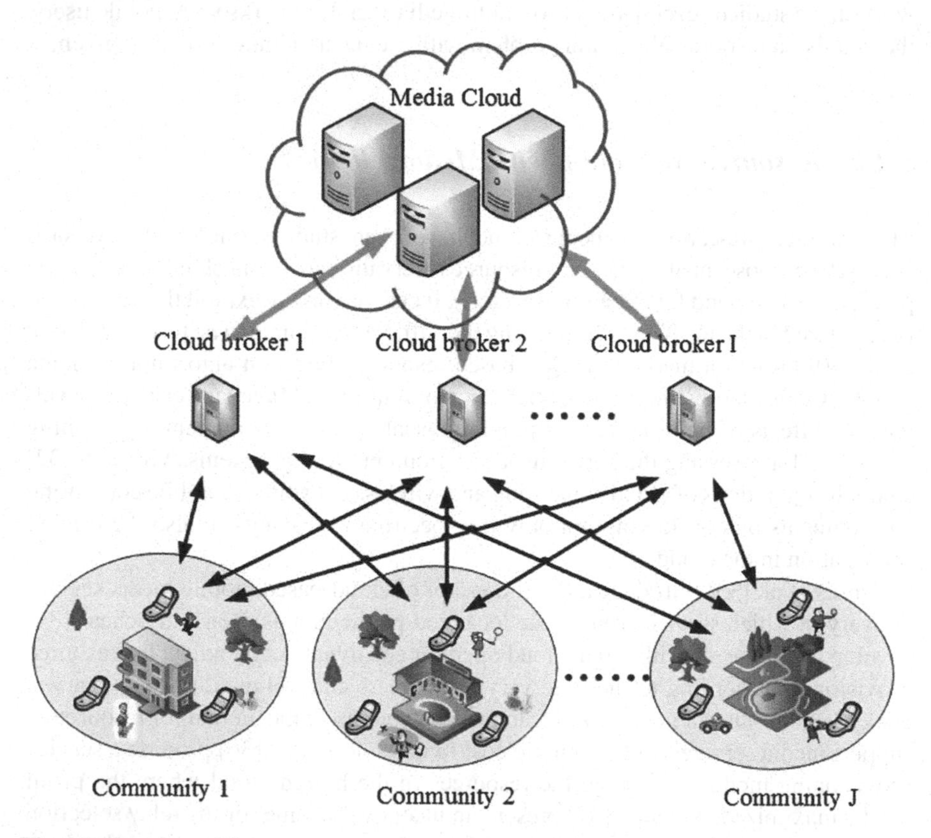

Fig. 5.1 System model

a community. In the community, users can select a broker to obtain the resource and observe others' strategies on the selections of brokers. The system model consists of the following components:

- *Users* With mobile devices, users can have the demands for media applications and send tasks to media cloud for processing [38, 39]. A social community is formed by a group of users, who have the similar interests, goals, or locations. Usually, users in the same community have social relations with each other, where a user can know the information of others. Let $\mathbb{J} = \{1, 2, \ldots, J\}$ denote the set of communities in the network, where the number of communities is J. The set of users in community j is denoted as $\mathbb{U}_j = \{u_{j,1}, u_{j,2}, \ldots, u_{j,k}, \ldots, u_{j,N_j}\}$ and there are N_j users in this community.
- *Cloud brokers* The set of cloud brokers is denoted as $\mathbb{I} = \{1, 2, \ldots, I\}$, where there are I cloud brokers in total. The cloud brokers are placed closely to users. Besides, the media cloud and cloud brokers are connected with high speed communication links. In practice, the cloud brokers [14] can be seen as the proxy between media cloud and users. The cloud brokers receive the media tasks from users and then buy the corresponding resources to process the tasks. The advantages of introducing cloud brokers are as follows. Firstly, due to the high speed communications between media cloud and cloud brokers, the service response time can be significantly reduced so that users can obtain the media services quickly. Secondly, for media cloud, as it directly connects cloud brokers and the number of brokers is less than users, media cloud can decrease the cost of access control and transmission.
- *Media cloud* Media cloud can provide virtual resources (computing, storage, and cloud service etc.) to users via cloud brokers. Based on [40], in this chapter the resource of media cloud can be described as the processing rate which media cloud can provide to deal with the multimedia tasks. Media cloud is responsible to process users' media tasks received from cloud brokers, and then return the corresponding results to users via broker with an allocated processing rate. We assume that the media cloud can totally provide B resource to users via brokers.

Based on [41], the broker in our system can have five modules, which are high speed communication module, wireless communication module, price decision-making module, task receiving module, and task delivering module. The high speed communication module is used to communicate with media cloud through the wired connection with high speed. The wireless communication module is used to communicate with model devices with wireless connection. The price decision-making module is to decide the price of resource to sell to users. In addition, the task receiving module is to sell resource to users and receive tasks from mobile devices. The task delivering module is to buy resource from cloud and deliver tasks to media cloud for processing.

Mobile devices can connect with brokers through wireless connection. Firstly, the brokers buy the resource from the media cloud after determining the price of resource. Then, users determine the optimal strategy on the resource demand. Next, users send the corresponding tasks to brokers with wireless communication. In addition, the brokers deliver the tasks to media cloud by wired module. At last, the results of tasks are delivered back to users.

Users can adjust and change their selection of broker based on the strategies of others in the community. Specifically, if a mobile user observes another's utility is larger than his utility, this user can change his selection. The selection of users can be modeled as the evolutionary game, where the result of game is that all mobile users in the community have the identical utility. Besides, users can check the brokers which are in the communication coverage and then determine one broker of them to connect.

5.3.2 QoE Model

QoE is to measure users' satisfactory when multimedia services are provided. In this chapter, due to the feature of media cloud, we present a QoE model by defining the obtained cloud processing rate of r, where $Q(r) = h(r)$. Moreover, based on [42], we assume that the QoE function has the following properties:

- $h(r)$ is positive;
- $h(r)$ is concave with respect to r;
- $h(r)$ is continuous and twice differentiable for r.

Note that user's QoE follows the logarithmic law and QoE function can be modeled in the logarithmic form for applications of multimedia task [43]. Therefore, we adopt the QoE model as the logarithmic function, which is defined as

$$Q(r) = h(r) = \begin{cases} q_{\max}, & r > r_{\max} \\ \alpha \log_2(\beta r), & r_{min} \leq r \leq r_{\max} \\ q_{\min}, & r < r_{\min} \end{cases}. \tag{5.1}$$

where α and β are two constant parameters. Both of them are positive and can be different for various types of applications. Parameter $q_{\max}$ is the maximum of QoE when user obtains the highest processing rate $r_{\max}$, and parameter $q_{\min}$ is the minimum of QoE when user obtains the lowest processing rate $r_{\min}$.

As QoE function is continuous, $r_{\min}$ can be obtained by

$$r_{\min} = \frac{2^{\frac{q_{\min}}{\alpha}}}{\beta}. \tag{5.2}$$

Similarly, $r_{\max}$ can be obtained by

$$r_{\max} = \frac{2^{\frac{q_{\max}}{\alpha}}}{\beta}. \tag{5.3}$$

5.3.3 *Design Goals*

Our design goals have two desirable objectives as follows: on one hand, user can obtain sufficient cloud resource from media cloud to achieve a satisfied QoE. On the other hand, media cloud can achieve the maximum profit with an optimal price of cloud price.

5.4 Problem Formulation

In this section, we propose a resource allocation among users, brokers, and media cloud. Firstly, the resource allocation framework is introduced. Then, the utility function of each player is defined. Finally, the structure of game is elaborated in detail.

5.4.1 *Resource Allocation Framework*

There are three parties in the resource allocation framework, which are media cloud, brokers, and users. Media cloud determines the price of cloud resource p, which denotes the unit of resource to be paid by brokers. Each broker i decides the amount of cloud resource E_i to buy, and then announces the price of the bought resource p_i to users. Each user decides a broker to connect for acquiring media service and obtaining the satisfied QoE. Due to the social features of MSNs, users in the same community are friends who can communicate with each other. Thus, in the same community, each user can know the information of others' connections and then compare the utility of his connection with others. If anyone's utility is better than his, he can change his connection.

The choices of three parties can mutually affect the decision of each other. The higher price of resource is determined by a media cloud, the less cloud resource that the brokers may decide to purchase, even this broker having heavy tasks. At the same time, if one broker announces a higher price to users, the user will connect other broker who announces lower price.

To allow the coordination among three parties to choose the proper parameter for enhancing performances, we propose a resource allocation framework with four steps. Firstly, the media cloud determines the price of its cloud resource and broadcasts it to brokers, aiming to acquire the revenue. There is a tradeoff between the revenue and the price for media cloud, e.g., if the price is too high, the demand of cloud resource will be reduced, which will affect the overall revenue. Secondly, after receiving the price of resource, each broker can buy a certain size of cloud resource from the media cloud to satisfy the demand of users who connect to this broker.

Thirdly, each broker decides the price of resource to achieve revenue from users. At last, each user will select a broker for connection and send his task to the broker.

5.4.2 Utility Functions

The utility functions of users To quantify the utility obtained from the resource, user utility considers the price of resource and the processing rate based on the acquired resource. According to the logarithmic function of allocated resource [44], the payoff of a user can be formulated as

$$s(r) = \varepsilon \log (1 + f(r). \tag{5.4}$$

where ε is a payoff parameter and $f(r)$ is the function of the acquired cloud resource from the connected broker.

Here, user in community j will select a proper broker to buy cloud resource, aiming to maximize his payoff with the least cost. Therefore, the utility function of a user in community j is defined as the difference between the payoff and cost on resource by

$$U_{i,j}(r_i) = s_{i,j}(r_i) - C_{i,j}(r_i), \tag{5.5}$$

where $s_{i,j}(r_i)$ denotes the payoff of a user in community j who connects broker i, and $C_{i,j}(r_i)$ is defined as the cost for buying the cloud resource from broker i. The payoff can be obtained by

$$s_{i,j}(r_i) = \varepsilon_{i,j} \log \left(1 + Q_{i,j}\left(\frac{r_i}{n_i}\right)\right), \tag{5.6}$$

Here, $Q_{i,j}\left(\frac{r_i}{n_i}\right)$ denotes the QoE of a user in community j who connects broker i. And broker i has bought r_i resource from media cloud. n_i is the number of users who connect broker i. As users in the same community may share the resource of broker i, users have the identical amount of resource when connecting broker i. Thus, the QoE of user j connecting broker i can be defined as

$$Q_{i,j} = \alpha_{i,j} \log_2 \left(\beta_{i,j} \frac{r_i}{n_i}\right). \tag{5.7}$$

Here $\alpha_{i,j}$ and $\beta_{i,j}$ are two constants of a user in community j who connects broker i, and they are related to media applications, which imply the sensitivity of a user on satisfaction of the obtained resource. For example, the sensitivity of watching video is higher than listening to the music with media cloud.

The cost for buying the cloud resource from broker i can be obtained by

$$C_{i,j}(r_i) = p_i, \tag{5.8}$$

where p_i is the price of resource. Therefore, the utility function of a user in community j who has the connection with broker i can be defined as

$$U_{i,j}(r_i) = \varepsilon_{i,j} \log\left(1 + \alpha_{i,j} \log_2\left(\beta_{i,j}\frac{r_i}{n_i}\right)\right) - p_i. \tag{5.9}$$

The objective of user is to achieve a large QoE with a cost as low as possible, in order to maximize its utility. Thus, the optimization problem for a user in community j connecting broker i can be formulated by

$$\max_{r_i} U_{i,j}(r_i). \tag{5.10}$$

$$s.t. \begin{cases} r_i & \geq 0 \\ \log_2\left(\beta_{i,j}\frac{r_i}{n_i}\right) & > 0 \end{cases}. \tag{5.11}$$

The utility functions of brokers For each broker, it provides the cloud resource for processing users' media tasks. The utility of broker is the revenue obtained from users minus the cost to buy cloud resource from media cloud. Thus, the utility of broker i can be defined as

$$U_i(p_i, E_i) = R_i(p_i) - C_i(E_i). \tag{5.12}$$

where $R_i(p_i)$ is the revenue through selling cloud resource from broker i to users, and $C_i(E_i)$ denotes the cost to obtain cloud resource from the media cloud.

We can obtain the revenue from selling the cloud resource by

$$R_i(p_i) = n_i p_i. \tag{5.13}$$

According to the pricing strategy of media cloud, the cost function can be denoted by

$$C_i(E_i) = p(D_i + E_i), \tag{5.14}$$

where p is the real-time price of cloud resource. D_i denotes the cloud resource to support the basic operation of broker i, it can be seen as the reserved resource which is provided to brokers by media cloud. E_i denotes the additional cloud resource to conduct the media task when broker i is busy. Therefore, the utility function of broker i is

$$U_i(p_i, E_i) = n_i p_i - p(D_i + E_i). \tag{5.15}$$

We assume that there is a discount when brokers obtain cloud resource from media cloud due to the transmission loss between media cloud and brokers. Thus,

$$r_i = \varsigma E_i, \tag{5.16}$$

where ς is the discount parameter.

The objective of broker is to achieve the revenue and reduce the cost as much as possible, for maximizing its utility. Thus, the optimization problem for broker i can be formulated as

$$\max_{p_i, E_i} U_i(p_i, E_i). \tag{5.17}$$

$$s.t. \begin{cases} r_i & = \varsigma E_i \\ p_i & \geq 0 \\ E_i & \geq 0 \end{cases}. \tag{5.18}$$

The utility function of media cloud By selling the cloud resource with a certain price to brokers, media cloud can obtain the corresponding revenue. In addition, the cost for processing media task should be also considered. Thus, the utility function of media cloud is defined as the difference between the revenue and the cost by

$$U_r(p) = R_r(p) - C_r. \tag{5.19}$$

where $R_r(p)$ denotes the revenue that cloud resource can obtain and C_r is the cost of media cloud for operation.

The revenue of media cloud by selling of cloud resource can be obtained by

$$R_r(p) = \sum_{i=1}^{I} p(D_i + E_i). \tag{5.20}$$

The cost for processing tasks is defined as

$$C_r = \sum_{i=1}^{I} c_r(D_i + E_i), \tag{5.21}$$

where c_r denotes the unite cost. Therefore, the utility function of cloud becomes

$$U_r(p) = \sum_{i=1}^{I} p(D_i + E_i) - \sum_{i=1}^{I} c_r(D_i + E_i). \tag{5.22}$$

The objective of media cloud is to achieve the large revenue by maximizing its utility. Thus, the optimization problem for media cloud can be formulated as

$$\max_{p} U_r(p). \tag{5.23}$$

$$s.t. p \geq 0. \tag{5.24}$$

5.4.3 *Four Stage Stackelberg Game*

We formulate the above problem as a four stage Stackelberg game, by considering the utility maximization of media cloud, brokers and users. In stage I, as a leader in the Stackelberg game, media cloud offers a real-time cloud resource price p to brokers. In stage II, as a follower in stage I, the broker decides the amount of cloud resource E_i based on the price offered by media cloud. Next, the broker acts as the leader in stage III, and offers the resource price to users. In stage IV, each user selects a proper broker to connect to acquire media service, based on the resource price and availability of cloud resource offered by the broker.

5.5 Analysis of the Proposed Four Stage Game

In this section, we analyze the proposed four-stage Stackelberg game, and obtain its Stackelberg equilibrium. Based on the above analysis, it is known that each stage's strategy may affect other stages' strategies. Therefore, we use the backward induction method to analyze the proposed game, as it can capture the sequential dependence of the decisions in each stage of the game.

5.5.1 *Evolution Game Among Users in Stage IV*

Communities are formed by groups of users with media service demands. Especially, users in the same community are friends of each other. They can communicate with each other, where the information can be exchanged among them. Therefore, users in the same community can observe others' decisions on the selection of brokers, and then adjust his strategy to be optimal. We propose an evolutionary game model to solve the broker selection problem. In the evolutionary game, users are the players of the game. The community represents a population in the game.

Replicator dynamic is crucial to analyze the evolution game to obtain the game equilibrium, where the utility of all users in a community are identical. And no player will change his current strategy because the rate of strategy change is zero. For community j, the proportion of users who select broker i to acquire media service becomes

$$x_{i,j} = \frac{n_{i,j}}{N_j}. \tag{5.25}$$

where $n_{i,j}$ is the number of users in community j to connect broker i, and N_j is the number of users in community j. We denote the state of community as the proportions of users to connect brokers. Thus, the state of community j can be obtained by

$$\mathbf{x}_j = \left[x_{1,j}, x_{2,j}, \ldots, x_{i,j}, \ldots, x_{I,j}\right]. \tag{5.26}$$

In the replicator dynamic, the share of a strategy in community grows at a rate which is directly proportional to the difference between the user's utility and the average utility. It can be denoted as

$$\dot{x}_{i,j}(t) = \lambda x_{i,j}(t)(U_{i,j}(t) - \tilde{U}_j(t)), \tag{5.27}$$

where λ is the multiplier of the difference between the user's utility and the average utility. $\tilde{U}_j(t)$ is the average utility of the entire community j. It can be calculated by

$$\tilde{U}_j(t) = \sum_{i=1}^{I} x_{i,j} U_{i,j}(t). \tag{5.28}$$

From the above, it can be obtained $\sum_{i=1}^{I} \dot{x}_{i,j}(t) = 0$. Therefore $\sum_{i=1}^{I} x_{i,j} = 1$ is satisfied during the broker selection process. Substituting (5.9) into (5.27), we have

$$\begin{aligned}\dot{x}_{i,j}(t) = \lambda x_{i,j}(t) &\left(\varepsilon_{i,j} \log\left(1 + \alpha_{i,j} \log_2\left(\beta_{i,j}\frac{r_i}{N_i}\right)\right)\right. \\ &\left. - p_i - \sum_{i=1}^{I} x_{i,j}\left(\varepsilon_{i,j} \log\left(1 + \alpha_{i,j} \log_2\left(\beta_{i,j}\frac{r_i}{N_i}\right)\right) - p_i\right)\right).\end{aligned} \tag{5.29}$$

We consider the evolutionary equilibrium as the solution to the broker selection game among users. An evolutionary equilibrium is a fixed point of the replicator dynamic. At the fixed point, which can be obtained numerically, the payoff of all users in community j are identical. In other words, since the rate of strategy adaptation is zero, the equilibrium can be obtained by solving

$$\dot{x}_{i,j}(t) = 0, \quad 1 \le i \le I, \quad 1 \le j \le J. \tag{5.30}$$

To evaluate the stability at the fixed point $x^*_{i,j}$, which is obtained by solving (5.30), the eigenvalues of the Jacobian matrix which is corresponding to the replicator dynamic needs to be evaluated. The fixed point is stable if each eigenvalue has a negative real part [45]. Here we have the evolutionary equilibrium for any community j as follows.

$$\mathbf{x}^*_j = \left(x^*_{1,j}, x^*_{2,j}, \ldots, x^*_{i,j}, \ldots, x^*_{I,j}\right). \tag{5.31}$$

5.5.2 Non-cooperative Game Among Brokers in Stage II and Stage III

Based on the result of the evolutionary game for users, the brokers compete with each other and choose the proper strategies on the price to obtain the maximum utilities. Thus, the non-cooperative game is introduced to model the competition among brokers, and the Nash equilibrium is considered as the solution to the game.

According to the price of the cloud resource determined by media cloud, each broker decides the amount of cloud resource to purchase and then determines the price of cloud resource to charge. Considering the competition, the utility of broker i can be defined as

$$U_i\left(E_i, p_i, \mathbf{E}_{-i}, \mathbf{p}_{-i}\right) = p_i \sum_{j=1}^{J} x_{i,j}^* N_j - p(D_i + E_i). \tag{5.32}$$

Here $\mathbf{E}_{-i}$ denotes the vectors of the resource size that brokers have, except broker i. $\mathbf{p}_{-i}$ means the price of cloud resource offered by brokers, except broker i.

The Nash equilibrium is considered as the solution of the game, where each broker has an optimal strategy to maximize the utility. In this case, we use the best response function of each broker to find Nash equilibrium, which is the best strategy of a broker based on others' best strategies. Therefore, when others' strategies are determined, the best response function of broker i can be defined by

$$\mathbb{B}\left(\mathbf{E}_{-i}, \mathbf{p}_{-i}\right) = \arg\max_{E_i, p_i} U_i(E_i, p_i, \mathbf{E}_{-i}, \mathbf{p}_{-i}). \tag{5.33}$$

Let $\mathbf{p}^* = (p_1^*, p_2^*, \ldots, p_I^*)$ and $\mathbf{E}^* = (E_1^*, E_2^*, \ldots, E_I^*)$ denote Nash equilibrium of the cloud resource price and the cloud resource size obtained from media cloud, respectively. The Nash equilibrium of the game can be obtained by solving

$$p_i^* = \mathbb{B}\left(\mathbf{E}_{-i}, \mathbf{p}_{-i}^*\right), \tag{5.34}$$

$$E_i^* = \mathbb{B}\left(\mathbf{E}_{-i}^*, \mathbf{p}_{-i}\right), \tag{5.35}$$

where $\mathbf{p}_{-i}^*$ and $\mathbf{E}_{-i}^*$ is the set of Nash equilibrium of brokers except broker i.

From the above analysis, a broker needs the strategy of other brokers and the equilibrium of the evolutionary game to obtain Nash equilibrium. However, this information may not be available in a practical broker system. Therefore, each broker can only employ the local information and users' demands to determine the prices and the cloud resource. Then, each broker should adjust its strategy in the direction of utility maximization. Therefore, broker i updates the price of cloud resource and the cloud resource size by

$$p_i(\tau+1) = p_i(\tau) + \omega_{i,p}\frac{\partial U_i(\mathbf{E}(\tau), \mathbf{p}(\tau))}{\partial p_i(\tau)}. \tag{5.36}$$

$$E_i(\tau+1) = E_i(\tau) + \omega_{i,E}\frac{\partial U_i(\mathbf{E}(\tau), \mathbf{p}(\tau))}{\partial E_i(\tau)}. \tag{5.37}$$

Here $p_i(\tau)$ and $E_i(\tau)$ are the price of cloud resource to sell and the size of cloud resource purchased from the media cloud. Both of them are determined by broker i at iteration τ. $\omega_{i,E}$ and $\omega_{i,p}$ are used to control the speed of adjustment on the cloud resource size and cloud resource price. The marginal payoff can be used to update the strategy for each broker [46]. It can be calculated by the variation in payoffs with a small variation φ (e.g., $\varphi = 10^{-4}$) as follows:

$$\begin{aligned}&\frac{\partial U_i(\mathbf{E}(\tau), \mathbf{p}(\tau))}{\partial p_i(\tau)}\\ &\approx \frac{U_i(\ldots, p_i(\tau)+\varphi, \ldots) - U_i(\ldots, p_i(\tau)-\varphi, \ldots)}{2\varphi}.\end{aligned} \tag{5.38}$$

$$\begin{aligned}&\frac{\partial U_i(\mathbf{E}(\tau), \mathbf{p}(\tau))}{\partial E_i(\tau)}\\ &\approx \frac{U_i(\ldots, E_i(\tau)+\varphi, \ldots) - U_i(\ldots, E_i(\tau)-\varphi, \ldots)}{2\varphi}.\end{aligned} \tag{5.39}$$

5.5.3 *Strategy of Media Cloud in Stage I*

The cloud resource can be sold to the broker to obtain revenue by media cloud. Thus, the media cloud hopes to choose a proper price of cloud resource to obtain the maximum utility. For media cloud, the optimization problem can be formulated as

$$p^* = \arg\max_p U_r(\mathbf{E}, p), \tag{5.40}$$

where p^* is the optimal strategy of media cloud on the price per cloud resource unit, $\mathbf{E} = [\mathbf{E}_1, \mathbf{E}_2, \ldots, \mathbf{E}_I]^T$ is the vector of cloud resource purchased by each broker.

Similar to the broker iteration, we also present a media cloud iteration to adjust the cloud resource price to obtain the maximum utility. The media cloud updates its price by

$$p(t+1) = p(t) + \omega_r\frac{\partial U_r(\mathbf{E}(t), p(t))}{\partial p(t)}, \tag{5.41}$$

where ω_r is used to control the speed of adjustment on the price of cloud resource price.

The marginal payoff can be calculated by

$$\frac{\partial U_r(\mathbf{E}(t), p(t))}{\partial p(t)} \approx \frac{U_r(\mathbf{E}(t), p(t)+\varphi) - U_r(\mathbf{E}(t), p(t)-\varphi)}{2\varphi}. \tag{5.42}$$

Here, when all users obtain the maximum utilities with the optimal strategies, the evolutionary game reaches the equilibrium. If someone tries to adjust his selection to connect the broker, the number of connection of this broker will become larger and the utilities of users in the same community to connect with this broker will decrease. If the equilibrium state is not Pareto efficiency, the utility of a user can be larger by adjusting strategy, where other users in the same community may imitate this selection to obtain higher utilities with the result that all users in the same community have the identical utility. In this case, the state in evolution game is not stable, which is not the equilibrium. Therefore, as the equilibrium can be obtained which is opposite to the above assumption, the equilibrium of evolution game in our work is Pareto efficiency. In addition, when the Stackelberg game reaches to the equilibrium, each broker or media cloud only has one optimal strategy. Therefore, each party can not adjust strategy to obtain higher utility when other two parties choose the optimal strategy. It also proves the Pareto efficiency of the proposed scheme.

5.5.4 Algorithm Design for Scheme Implementation

Based on the above analysis of four-stage Stackelberg game, we present an iteration algorithm to implement our scheme. For the cloud resource, it can update its cloud resource price to obtain a maximum utility and then announce this price to all brokers. Since the media cloud is not aware of the duration of each adjustment, the media cloud sets a waiting time $T_{w,mc}$ for the next strategy update. Similarly, as each broker is not aware of the duration of each evolution, it sets a waiting $T_{w,b}$ for the next strategy to update the size of the purchased cloud resource and the price to charge users. In the evolutionary game, each user randomly selects a broker to connect initially, and then changes his strategy to maximize his own utility. If a users utility is lower than the average utility of his community, this user may change his connection with a probability, denoted by

$$\theta = \frac{\tilde{U}_j - U_{i,j}}{\tilde{U}_j}, \tag{5.43}$$

where $\tilde{U}_j$ is the average utility of community j. When all users in the same community obtain an equal utility, the evolution will be completed. We present the algorithm by Algorithm 1.

Algorithm 1: Resource allocation iteration algorithm

1: Initially, the media cloud announces the price $p(0)$ to all brokers.
2: **Repeat**
3: **while** $t \leq T_{w,mc}$ **do**
4: Each broker randomly determines the size of leased resource E_i and the price p_i.
5: **Repeat.**
6: **while** $t \leq T_{w,b}$ **do**
7: Each user randomly makes connection.
8: **Repeat.**
9: Compute each user's utility by (5.22).
10: Exchange connection information with each other in the community.
11: Calculate the average utility $\widetilde{U}_j$ by (5.28).
12: **if** $\widetilde{U}_j > U_{i,j}$ **then**
13: Change the connection with probability θ.
14: **else**
15: Maintain the connection.
16: **end if**
17: **Until** all users in the same community have the equal utility.
18: **end while**
19: Update cloud resource size $E_i(\tau)$ and the cloud resource price $p_i(\tau)$ by equations (5.36)-(5.39).
20: $\tau = \tau + 1$
21: **Until** E_i and p_i are both unchanged.
22: **end while**
23: Update the price $p(t)$ by Eqs. (5.41), (5.42).
24: $t = t + 1$
25: **Until** p is unchanged.

5.6 Performance Evaluation

In this section, we evaluate the performance of the proposed game-based cloud resource allocation.

5.6.1 Simulation Setup

In the simulation, there is a media cloud to lease cloud resource to brokers. The total size of cloud resource is $B = 100\,\text{k}$ which denotes that the cloud can process 100×10^3 tasks per unit time. In addition, the MSN has two brokers and two communities. In each community, there are 20 users. The media cloud waits $T_{w,mc} = 500$ for the next strategy update and each broker waits $T_{w,b} = 100$ for the next generation strategy update. The speed of adjustment for each broker on the bought resource size and price are $w_{i,p} = 0.1$ and $w_{i,E} = 1$, respectively. The speed of adjustment for media cloud on price of resource is $w_r = 0.01$. The detailed values of parameters in this simulation are listed in Table 5.1.

Table 5.1 Simulation parameters

Parameter	Value
J: the number of communities	2
N_1, N_2: the number of users in community 1 and community 2	20, 20
I: the number of cloud brokers in the network	2
B: the maximum of resource can be allocated by media cloud	100k (tasks/s)
$\{\alpha_{1,j}, \beta_{1,j}\}, j \in \{1, 2\}$: the parameters on QoE of users in community 1	{2, 10}
$\{\alpha_{2,j}, \beta_{2,j}\}, j \in \{1, 2\}$: the parameters on QoE of users in community 2	{1.5, 10}
$\{\varepsilon_{1,j}, \varepsilon_{2,j}\}$: the payoff parameters of mobile users in community 1 and community 2	{2, 2}
D_i: the reserved resource provided to brokers	0
ς: the discount parameter	1
$T_{w,mc}$: the waiting time for strategy adjustment of media cloud	500
$T_{w,b}$: the waiting time for strategy adjustment of a broker	100

5.6.2 Numerical Results

Firstly, we study the evolutionary behavior of users. We set the cloud resource size and the price of two brokers as $E_1 = E_2 = 20$ and $p_1 = 0.1$, $p_2 = 0.5$, respectively. Figure 5.2 shows the convergence of the evolutionary behavior of users when the initial state of community is $(x_{1,1}, x_{1,2}) = (0.4, 0.4)$. From Fig. 5.2, we can observe that both of utilities of users in community 1 and community 2 are converged to be optimal with several iteration steps. In addition, it can be known that the utilities of all users in both community 1 and community 2 are nearly identical.

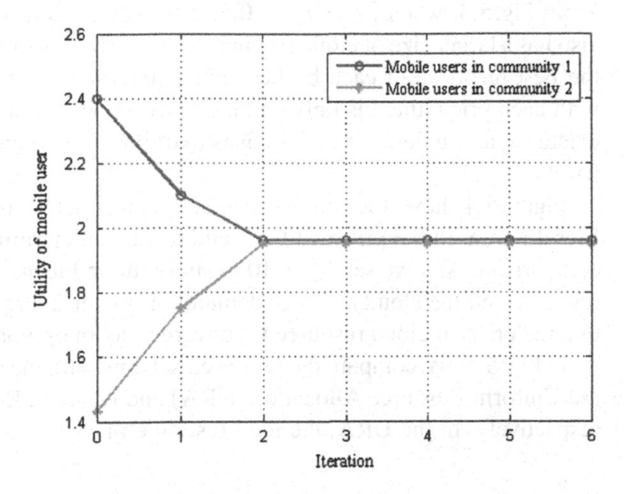

Fig. 5.2 Convergence of the evolution among users to the equilibrium

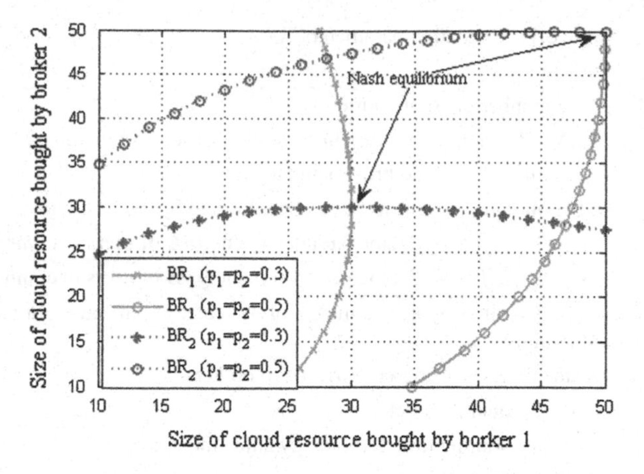

Fig. 5.3 Best response of each broker on the size of cloud resource purchased from the media cloud. BR_1 and BR_2 represent the best response functions of broker 1 and broker 2

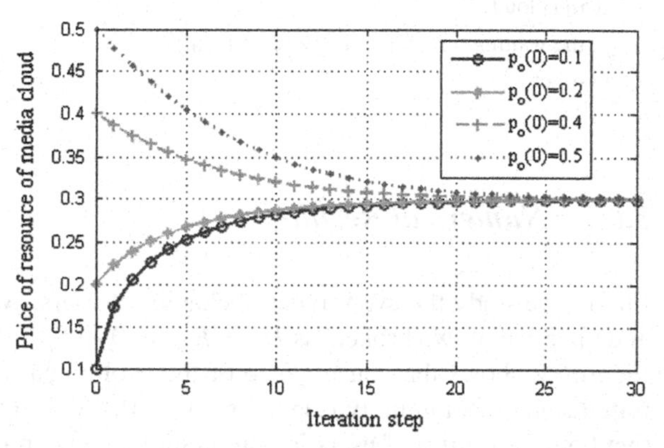

Fig. 5.4 Price of cloud resource determined by the media cloud versus iteration step. The initial prices $p_o(0)$ are 0.1, 0.2, 0.4, 0.5, respectively

Figure 5.3 shows the best response of each broker on the size of cloud resource purchased from media cloud. We set $p_0 = 0.1$ and choose two types of cloud service price for comparison, which are $p_1 = p_2 = 0.3$ and $p_1 = p_2 = 0.5$, respectively. From Fig. 5.3, when $p_1 = p_2 = 0.5$, if broker 2 has more cloud resource, broker 1 also has a larger size of cloud resource to be the best response. When $p_1 = p_2 = 0.3$, the best response of each broker firstly increases and then decreases. In addition, with each price, there is only one intersection point in Fig. 5.3. It demonstrates the existence and uniqueness of the Nash equilibrium when the cloud service price is fixed.

Figure 5.4 shows the convergence of the price determined by media cloud. We set four different initial prices of the cloud resource determined by the media cloud for comparison. And we set $E_2 = 10$ to study the influence between the price of cloud resource and the cloud resource demand of broker 2. From Fig. 5.4, we can observe that the price of cloud resource is converged to an optimal price with several steps.

In Fig. 5.5, we compare the proposed scheme with the existing approaches, which are Uniform Resource Allocation (URA) and Random Resource Allocation (RRA), respectively. In the URA, the total resource of media cloud is uniformly allocated

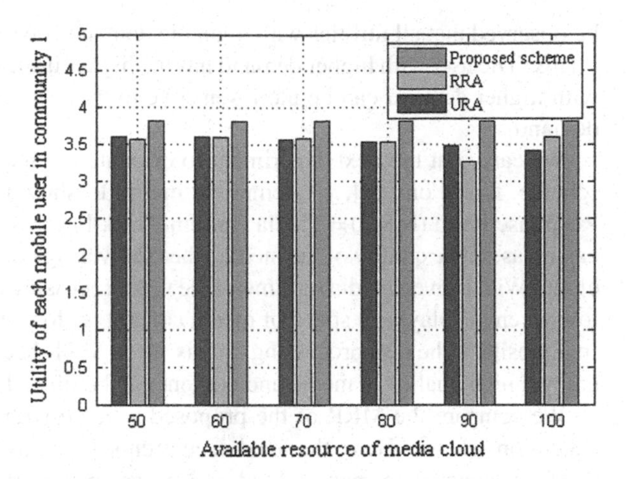

Fig. 5.5 Utility of each user in community 1, where the initial price of the resource determined by the media cloud is $p(0) = 0.5$

to all users in the network. In the RRA, each user can obtain cloud resource from the media cloud randomly. From Fig. 5.5, it can be known that the proposed scheme outperforms the other two existing approaches, where user can obtain the best utility. In the URA, as the cloud resource is uniformly allocated to users, too much resource may be allocated to someone whose demand is low, while users who need more resource can not obtain enough resource. In RRA, as the resource is allocated to users randomly, users cannot obtain the resource according to their needs. In the proposed scheme, users can obtain their wanted resource according to their demands. Furthermore, with the theoretical game model, the price gradually tends to be reasonable. Thus, all parties can possibly obtain the maximum utilities.

To test the performance with dynamical demands, Fig. 5.6 shows the utility of a user in community 1 when the value of in community 1 is changed from 2 to 3.5, which shows the variation of a user's resource demand [47]. From Fig. 5.6, it can

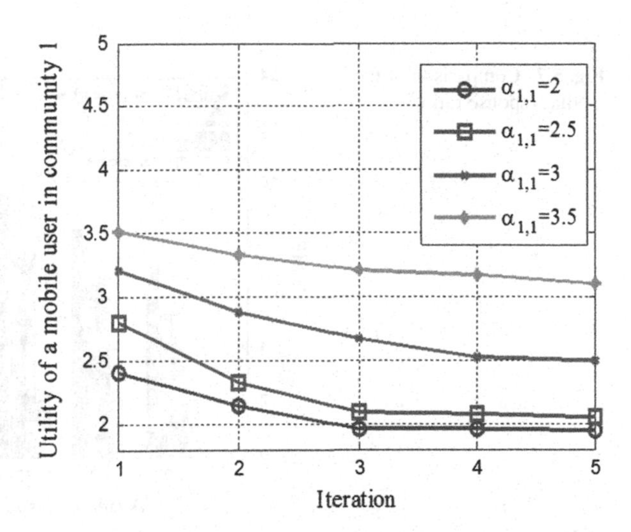

Fig. 5.6 The utility of mobile user in community when the demand is changed

be observed that all utilities with dynamic demands decrease and reach to the stable finally. The user with higher demand has the higher utility. The reason is that the user with higher demand can be more sensitive to the resource than the one with lower demand.

We carry out the next experiment to evaluate the media quality of the proposed scheme. Based on [48], we define the metric to show the media quality as Media Response Ratio (MRR) = Media Runtime/Task Processing Time. The above metric can measure the quality of the media when delivering content to users through media cloud. With a given media runtime, if the task processing time is long, MRR becomes low when the playback speed of media content is slow and the media may be stunk. In opposite, if the task processing time is short, MRR becomes large where users can enjoy a high quality of media and content can be played fluently.

We compare the MRR of the proposed scheme with RRA, URA, and the local execution scheme. Here, the local execution scheme means that the mobile device does not connect to media cloud and processes the media data on local device. According to [48], in the experiment the file size of media is determined as 307MB and the runtime is 1291 s. Without the cloud, the task process rate of local device is 500 tasks/s. For the task process rate of the proposed scheme, it is decided by the proposed algorithm. The task process rate of RRA and URA are determined randomly and uniformly, respectively. From Fig. 5.7, we can see that the proposed scheme can achieve the highest MRR compared to other schemes. The reason is that users can adjust the strategy to achieve the maximum revenue based on the social features in the community.

In the above experiments, it can be known that all users can choose the best strategies to obtain the optimal utility. Each broker can determine its optimal strategy on cloud service price and size to obtain the maximum utility. The price of cloud resource determined by media cloud is converged to the optimal. Therefore, the proposed resource allocation scheme is converged and the Stackelberg equilibrium exists. More details can be found in [49].

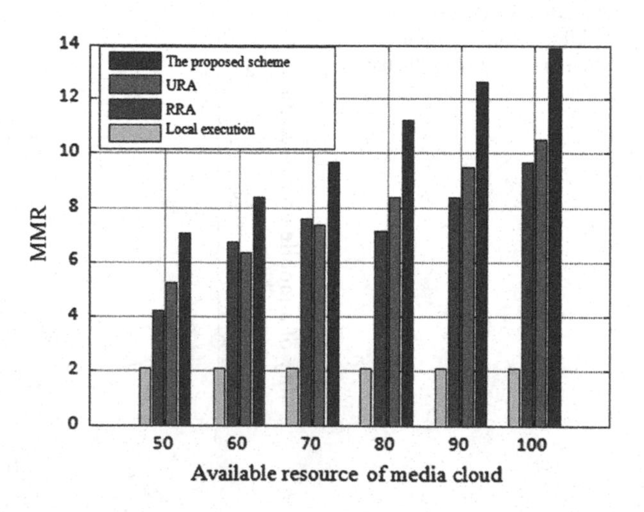

Fig. 5.7 Comparison of the media response ratio

5.7 Summary

In this chapter, based on the competition among media cloud, brokers and users, we have presented a resource allocation scheme for users to achieve satisfied QoE with media cloud. In the proposed scheme, the media cloud can determine a certain price to lease cloud resource to brokers. Each broker can determine the size of cloud resource to buy and then provide the cloud resource to users at certain price. The user can adjust his strategy to decide his connecting broker. The resource allocation problem has been formulated as a four-stage Stackelberg game. Through the backward induction method, we have proposed an iterative algorithm to obtain the Stackelberg equilibrium to implement the proposed scheme. Simulation results have been presented to demonstrate the performance of the proposal.

References

1. K. Zheng, Z. Yang, K. Zhang, P. Chatzimisios, K. Yang, W. Xiang, Big data-driven optimization for mobile networks toward 5G. IEEE Netw. **30**(1), 44–51 (2016)
2. Z. Su, Q. Xu, Content distribution over content centric mobile social networks in 5G. IEEE Commun. Mag. **53**(6), 66–72 (2015)
3. Z. Su, Q. Xu, H. Zhu, Y. Wang, A novel design for content delivery over software defined mobile social networks. IEEE Netw. **29**(4), 62–67 (2015)
4. C. Forecast, Cisco visual networking index: global mobile data traffic forecast update 2009–2014, in *Cisco Public Information*, vol. 9 (2010)
5. Y. Wu, S. Deng, H. Huang, Information propagation through opportunistic communication in mobile social networks. Mob. Netw. Appl. **17**(6), 773–781 (2012)
6. H. Zhu, S. Du, M. Li, Z. Gao, Fairness-aware and privacy-preserving friend matching protocol in mobile social networks. IEEE Trans. Emerg. Top. Comput. **1**(1), 192–200 (2013)
7. M. Li, S. Yu, N. Cao, W. Lou, Privacy-preserving distributed profile matching in proximity-based mobile social networks. IEEE Trans. Wirel. Commun. **12**(5), 2024–2033 (2013)
8. D. Adu-Gyamfi, Y. Wang, F. Zhang, M. Domenic, I. Memon, Y. Gustav, Modeling the spreading behavior of passive worms in mobile social networks, in *Proceedings of the ICIII*, vol. 1 (Xi'an, 2013), pp. 380–383
9. K. Zhang, X. Liang, R. Lu, X. Shen, PIF: a personalized fine-grained spam filtering scheme with privacy preservation in mobile social networks. IEEE Trans. Comput. Soc. Syst. **2**(3), 41–52 (2015)
10. P. Costa, C. Mascolo, M. Musolesi, G. Picco, Socially-aware routing for publish-subscribe in delay-tolerant mobile ad hoc networks. IEEE J. Sel. Areas Commun. **26**(5), 748–760 (2008)
11. H. Sun, C. Wu, Epidemic forwarding in mobile social networks, in *Proceedings of the IEEE ICC* (Ottawa, 2012), pp. 1421–1425
12. Y. Wu, S. Deng, H. Huang, Hop limited epidemic-like information spreading in mobile social networks with selfish nodes. J. Phys. A Math. Theor. **46**(26), 1–14 (2013)
13. Y. Wu, C. Wu, B. Li, L. Zhang, Z. Li, F. Lau, Scaling social media applications into geo-distributed clouds. IEEE/ACM Trans. Netw. (TON) **23**(3), 689–702 (2015)
14. X. Qiu, C. Wu, H. Li, Z. Li, F. Lau, Federated private clouds via broker's marketplace: a stackelberg-game perspective, in *Proceedings of the IEEE ICCC* (Alaska, 2014), pp. 296–303
15. Z. Su, Y. Hui, S. Guo, D2d based content delivery with parked vehicles in vehicular social networks, IEEE Wirel. Commun. **23**(8), 90–95 (2016)
16. N. Yu, Q. Han, Context-aware communities and their impact on information influence in mobile social networks, in *Proceedings of the IEEE PERCOM* (Lugano, 2012), pp. 131–136

17. M. Xiao, J. Wu, L. Huang, Community-home-based multi-copy routing in mobile social networks. IEEE Trans. Parallel Distrib. Syst. **13**(7), 3978–3900 (2014)
18. S. Zhan, S. Chang, Design of truthful double auction for dynamic spectrum sharing, in *Proceedings of the IEEE DYSPAN* (Virginia, 2014), pp. 439–448
19. I. Stanojev, A. Yener, Relay selection for flexible multihop communication via competitive spectrum leasing, in *Proceedings of the IEEE ICC* (Budapest, 2013), pp. 5495–5499
20. Y. Chang, H. Liu, L. Chou, Y. Chen, H. Shin, A general architecture of mobile social network services, in *Proceedings of the CIT* (Gyeongju, 2007), pp. 151–156
21. J. Wu, Y. Wang, Social feature-based multi-path routing in delay tolerant networks, in *Proceedings of the IEEE INFOCOM* (Florida, 2012), pp. 1368–1376
22. Y. Wang, J. Wu, W. Yang, Cloud-based multicasting with feedback in mobile social networks. IEEE Trans. Wirel. Commun. **12**(12), 6043–6053 (2013)
23. Z. Lu, Y. Wen, G. Cao, Information diffusion in mobile social networks: the speed perspective, in *Proceedings of the IEEE INFOCOM* (Toronto, 2014), pp. 1932–1940
24. Z. Li, C. Wang, S. Yang, C. Jiang, I. Stojmenovic, Improving data forwarding in mobile social networks with infrastructure support: a space-crossing community approach, in *Proceedings of the IEEE INFOCOM* (Toronto, 2014), pp. 1941–1949
25. W. Yin, T. Mei, C. Chen, S. Li, Socialized mobile photography: learning to photograph with social context via mobile devices. IEEE Trans. Multimed. **16**(1), 184–200 (2014)
26. J. Ren, Y. Zhang, K. Zhang, X. Shen, Exploiting mobile crowdsourcing for pervasive cloud services: challenges and solutions. IEEE Commun. Mag. **53**(3), 98–105 (2015)
27. X. Wang, M. Chen, T. Kwon, L. Yang, V. Leung, AMES-cloud: a framework of adaptive mobile video streaming and efficient social video sharing in the clouds. IEEE Trans. Multimed. **15**(4), 811–820 (2013)
28. Q. Xu, Z. Su, K. Zhang, P. Ren, X. Shen, Epidemic information dissemination in mobile social networks with opportunistic links. IEEE Trans. Emerg. Top. Comput. **3**(3), 399–409 (2015)
29. A. Alasaad, K. Shafiee, H. Behairy, V. Leung, Innovative schemes for resource allocation in the cloud for media streaming applications. IEEE Trans. Parallel Distrib. Syst. **26**(4), 1021–1033 (2015)
30. B. Hong, Y. Zhai, R. Tang, Y. Feng, A resources allocation algorithm based on media task qos in cloud computing, in *Proceedings of the IEEE of ICSESS* (Beijing, 2013), pp. 841–844
31. T. Magedanz, F. Schreiner, QoS-aware multi-cloud brokering for NGN services: tangible benefits of elastic resource allocation mechanisms, in *Proceedings of the IEEE of ICCE* (Danang, 2014), pp. 168–173
32. Z. Yin, F. Yu, S. Bu, Z. Han, Joint cloud and wireless networks operations in mobile cloud computing environments with telecom operator cloud. IEEE Trans. Wirel. Commun. **14**(7), 4020–4033 (2015)
33. F. Sardis, G. Mapp, J. Loo, M. Aiash, A. Vinel, On the investigation of cloud-based mobile media environments with service-populating and QoS-aware mechanisms. IEEE Trans. Multimed. **15**(4), 769–777 (2013)
34. S. Ren, M. Schaar, Efficient resource provisioning and rate selection for stream mining in a community cloud. IEEE Trans. Multimed. **15**(4), 723–734 (2013)
35. V. Aggarwal, V. Gopalakrishnan, R. Jana, K. Ramakrishnan, V. Vaishampayan, Optimizing cloud resources for delivering iptv services through virtualization. IEEE Trans. Multimed. **15**(4), 789–801 (2013)
36. P. Lu, Q. Sun, K. Wu, Z. Zhu, Distributed online hybrid cloud management for profit-driven multimedia cloud computing. IEEE Trans. Multimed. **17**(8), 1297–1308 (2015)
37. Q. Xu, Z. Su, S. Guo, A game theoretical incentive scheme for relay selection services in mobile social networks. IEEE Trans. Veh. Technol. **65**(8), 6692–6702 (2016)
38. D. Niyato, P. Wang, E. Hossain, W. Saad, Z. Han, Game theoretic modeling of cooperation among service providers in mobile cloud computing environments, in *Proceedings of the of IEEE WCNC* (Shanghai, 2012), pp. 3128–3133
39. K. Chard, S. Caton, O. Rana, K. Bubendorfer, Social cloud: cloud computing in social networks. IEEE CLOUD, **10**, 99–106 (2010)

40. X. Nan, Y. He, L. Guan, Optimal resource allocation for multimedia cloud in priority service scheme, in *Proceedings of the IEEE SCS* (Seoul, 2012), pp. 1111–1114
41. S. Kiani, M. Knappmeyery, N. Baker, B. Moltchanov, A federated broker architecture for large scale context dissemination, in *Proceedings of the IEEE CIT* (Bradford, 2010), pp. 2964–2969
42. W. Zhang, Y. Wen, Z. Chen, A. Khisti, QoE-driven cache management for http adaptive bit rate streaming over wireless networks. IEEE Trans. Multimed. **15**(6), 1431–1445 (2013)
43. P. Li, Y. Wang, W. Zhang, Y. Huang, QoE-oriented two-stage resource allocation in femtocell networks, in *Proceedings of the IEEE VTC* (Vancourver, 2014), pp. 1–5
44. M. Andrews, J. Cao, J. McGowan, Measuring human satisfaction in data networks. in *Proceedings of the IEEE INFOCOM* (2006)
45. Y. Kuznetsov, Elements of applied bifurcation theory **112**, XXII, 632 (Springer-Verlag New York, 2004). doi:10.1007/978-1-4757-3978-7
46. D. Niyato, E. Hossain, Z. Han, Dynamics of multiple-seller and multiple-buyer spectrum trading in cognitive radio networks: a game-theoretic modeling approach. IEEE Trans. Mobile Comput. **8**(8), 1009–1022 (2009)
47. Z. Su, Q. Xu, K. Zhang, K. Yang, X. Shen, Dynamic bandwidth allocation in mobile social networks with multiple homing access, in *Proceedings of the WCSP* (Nanjing, 2015), pp. 1–6
48. S. Kim, K. Kim, C. Lee, W. Ro, Offloading of media transcoding for high-quality multimedia services. IEEE Trans. Consum. Electron. **58**(2), 691–699 (2012)
49. Z. Su, Q. Xu, M. Fei, M. Dong, Game theoretic resource allocation in media cloud with mobile social users. IEEE Trans. Multimed. **18**(8), 1650–1660 (2016)

Chapter 6
Optimization of Heterogeneous MSN Architecture

In this chapter, we discuss the modeling and optimization for heterogeneous MSN architecture to deliver social contents. With the small cells and macro cells, we present a heterogeneous architecture where the content can be managed in a content-centric mode. Based on the availability of cached replicas in the content store of content centric nodes, a novel caching algorithm is proposed to replace replicas in order to efficiently use the cache capacity of the content store. In addition, simulation results show the efficiency of the proposed scheme.

6.1 Network Architecture for Heterogeneous MSNs

With the advance of network technologies and the innovation of mobile services [1–3], many efforts have been given by both academia and industry to design the fifth generation (5G) mobile networks [4, 5]. Some related projects such as 5GNOW [6] and 20BAH [7] are carried out in academia, and there are also many standardization activities in industry.

Among these research activities, MSNs keep attracting much attention [8], where the MSNs have been developed rapidly and millions of users can interact with each other to exchange content [9–12]. Besides, since 5G can make it possible for users to experience more emerging multimedia services including wearable mobile communications, augmented reality applications, etc., it can be predicted that MSNs will be one of the most important network paradigms in 5G.

However, compared with other conventional networks, MSNs pose some new challenges for development in 5G. First, the amount of content requests in MSNs is greater than others such that a higher delay is caused and it is hard to satisfy the requirements in 5G. Second, the replicas of one content may be stored at different sites. As these replicas are managed according to their different locations in the current networks, the overhead to manage these replicas on different sites incurs huge operating costs. In addition, with the rapid development of the Internet of things (IoT), connected vehicles, etc., various types of content need to be delivered

Z. Su et al., *Modeling and Optimization for Mobile Social Networks*,
DOI 10.1007/978-3-319-47922-4_6

efficiently while users are in motion [13]. New consideration to mobility should be given when designing MSNs in 5G.

In this chapter, we propose content centric based mobile social networks to resolve the above problems. First, in the content centric networks (CCNs) [14], content is delivered based on the interest in it instead of sending the conventional requesting message. In MSNs, the communities are organized by users with common interests. If the content is delivered by interest, delivering one content in a community may satisfy multiple users who have the same interest. Therefore, the total number of requesting messages in MSNs can be reduced. Second, in CCNs replicas are not controlled by their locations specified by IP addresses. Instead, the content is recognized by its content ID. This makes it possible to manage different replicas of the same content with the ID of the original content, and so the overhead to control these replicas can be reduced at the same time. Third, there are some CCN nodes in CCNs that cache some replicas of frequently used content within the coverage area of small cells. Therefore, when users keep moving through the coverage area of different cells, they can obtain the interested contents from the replicas cached in CCN nodes in small cells, without contacting the far away server.

To realize the above content-centric MSNs both content distribution using virtualization and related standardization are needed. On one hand, the construction of virtual CCNs has been studied for content distribution [15]. However, the network architecture and process of mobile social content delivery have not been mentioned. On the other hand, the IETF has carried out standardization activities to discuss the protocol for CCNs [16]. Although there are some standardized approaches such as the structure of the identifier and the status of the path, the related analytical models and algorithms need to be studied further, especially for content centric mobile social networks.

Therefore, in this chapter, we outline the delivery of mobile content based on content centric mechanism in MSNs. First, we present a heterogeneous network structure of content centric MSNs, which consist of users, communities, CCN nodes, small cells, and macro cells. Then, the process of content delivery based on the interest in content among users in communities is shown. Next, according to the availability of cached replicas in the content store of CCN nodes, we propose a novel caching scheme algorithm to replace replicas to efficiently use the cache capacity of the content store. In addition, experiment results are given to verify the proposed scheme.

6.2 Mobile Content Delivery

6.2.1 Requirement of Mobile Communication

The rapid increase of mobile traffic has placed a huge burden on current mobile networks [17], where the volume of content, the population of users, and types of mobile devices all keep growing quickly. Due to limited resources including power, spectrum, etc., the related analysis shows that performance expectations cannot always be satisfied by various mobile network applications [18–21]. Besides, recent mobile networks are expected to support new emerging applications, such as tactile Internet, augmented reality, wearable devices, and smart mobile social communication, etc. The existing networks are not reliable enough to meet the performance requirements of these new applications. Accordingly, the efficient design of 5G mobile networks is needed.

According to the standards of 5G mobile networks, there are some detailed requirements related to capacity, data rate, latency, connectivity, operating cost, and QoE provisioning, respectively [4]. For example, the aggregate data rate needs to be increased by 1000 times above the rate in 4G; the edge rate should be improved from 100 Mbps to 1 Gbps; the round time latency is expected to be 1 ms while it is 15 ms in 4G [5].

Nowadays, there are massive efforts to design 5G mobile networks for the year 2020 and beyond. Among these efforts, as one of the most popular and practical network paradigms to provide content services, the design of MSNs is an important issue in 5G.

6.2.2 Advance of Network Architecture for Mobile Social Contents

Different from the conventional network paradigms such as the client-server based structure, users in MSNs do not always need to contact far away servers to request content. Instead, they can directly communicate with each other to share content by short range wireless interfaces with peer to peer opportunistic links.

One of the most important features in MSNs is that users may have different interests in different content, and users with common interests can form a community based on their social ties. In the community, the content can be shared and distributed among these users. Small cells, which are covered by low power cell towers installed by the operator, are responsible for providing communication within the community. Compared with small cells, a macro cell is usually covered by the traditional base stations. The macro cell can provide communication within a large coverage area about a few kilometers. With cooperation among the macro cell, small cells, communities, and users, mobile content can be delivered in MSNs in 5G, with the main structure shown in Fig. 6.1. However, there are new challenges to deliver mobile content in 5G.

6.2.3 Challenges in Mobile Content Delivery in MSNs

How to Control a Huge Amount of Content Requests? In MSNs, users cannot only produce their own content, but also request content from others. For example, a

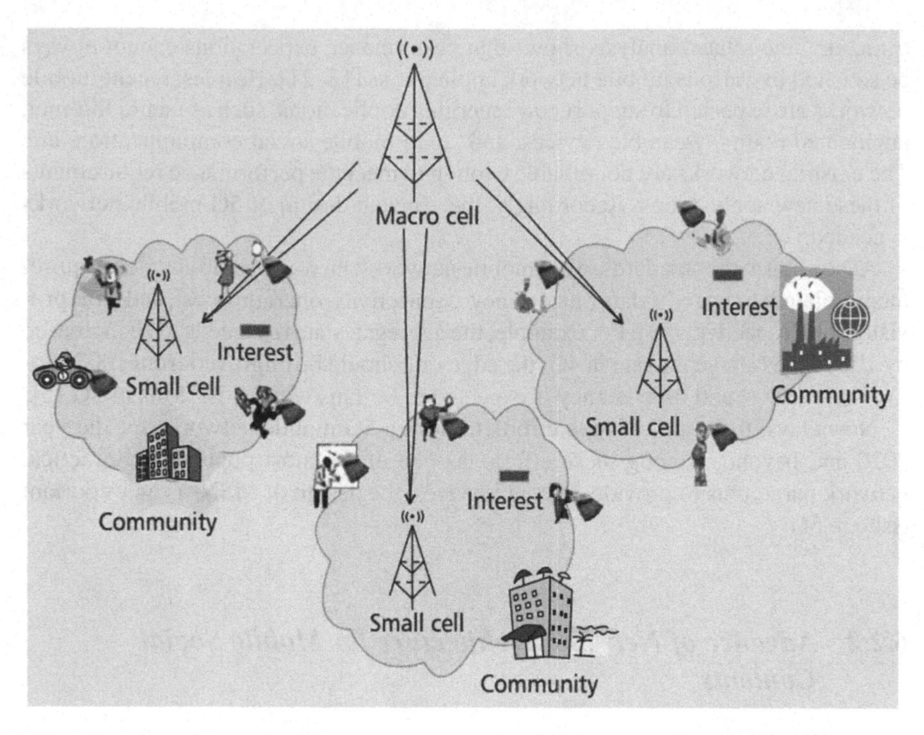

Fig. 6.1 Mobile content delivery in 5G

mobile user may update and publish his own content frequently, and can also subscribe to other people's content according to his interests and social ties. Compared with conventional networks, the amount of content requests in MSNs is not of the same order of magnitude. In particular, most requests are sent according to the interests of users. Since there is an obvious relationship based on interests among content, users, and communities, the requests should be controlled with a consideration of interests.

How to Control a Large Number of Replicas of Content? When a user publishes his own content, this content may be subscribed to by other users and then stored on different sites. Therefore, for the same content there will be a large number of its replicas, which incurs a high operating cost to manage these replicas. For example, when the producer of original content wants to update this content or publish some related information to this content, it is hard to recognize which replicas of his content are on which sites. Furthermore, due to mobility, it becomes even harder to control these replicas. Different from the conventional way of controlling replicas based on the location specified by an IP address, a new approach is required to manage replicas based on their own identity, e.g, its ID or name.

How to Control Load Balance? Compared with wired networks, both macro cells and small cells have their own limited capacities to deliver content. With the increase in the amount of content and the population of users, it is expected that network traffic can be distributed. Instead of forwarding all of the requests to the site that stores the original content, some content can be cached and placed within the small cell by different strategies to provide content with a possible load balance.

6.3 Content Centric Mobile Social Networks

6.3.1 Content Centric Networks

To develop MSNs in 5G, a new network structure is required, since the existing network architecture was designed based on the assumption that content delivery relies on the IP address of the content. However, with the rapid spread of multimedia content, in addition to the conventional server with a designated IP address storing the content, different users moving on different sites may also produce and consume content. Nowadays, content has less and less relation with where it is stored. For example, a user may publish the same content to two different communities located in different small cells. Although the content is the same, the existing networks manage it as two different contents due to the different locations. Therefore, the traditional approach to recognize and manage content based on the location specified by IP address will not be efficient now. In fact, current Internet users do not care where the content is from, they care more about what this content is. To face this new challenge, content centric networks (CCNs) [14–16, 22, 23] have been proposed as a new architecture for future networks to replace current networks.

As shown in Fig. 6.2, in CCNs there are many CCN nodes to store replicas of original content. The request for content is not sent to the IP address where this content is originally stored. Instead, an interest for the content is published over the CCNs; if a CCN node has the content where this interest refers, the replica of the content in this CCN node will be sent directly to the user who sent the interest. The interest in this content can be shown by a hierarchical name such as content ID. By sending the interest to the CCN node, the current IP address becomes unimportant, as the content requester need not know where the content was originally stored.

In the above CCN node, the content store (CS) is working with a pending interest table (PIT) and a forwarding information base (FIB). The content store caches the replicas of content. If a user sends an interest and the replica of this content is available in the content store, this replica will be provided to the user directly. Otherwise, the CCN node will check whether there is a pending interest for this content in the pending interest table. If there is still no matching interest, the CCN node will check the forwarding information base to wait for the wanted content fetched by a suggested delivering path. With cooperation among the content store, the pending interest table, and the forwarding information base, the content can be delivered according to the interest.

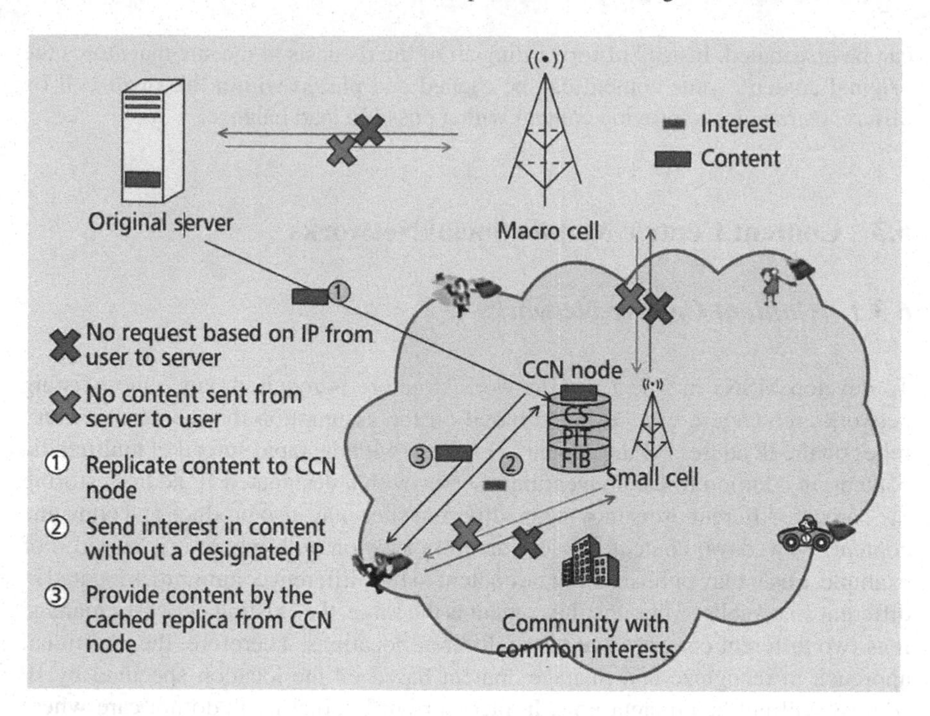

Fig. 6.2 Network architecture of content centric networks

6.3.2 Content Centric Mobile Social Networks

We propose the content centric mechanism as an efficient solution for mobile content delivery in MSNs in 5G for the following reasons.

Community with Common Interest versus Interest Based Content In MSNs, the community consists of users with common interests. As the content in CCNs is also delivered based on the interest, it is natural that multiple users can be satisfied with one interest. As shown in Fig. 6.3, when users in the community subscribe to content with the same interest, the original server only needs to send the content once to the small cell. Then the small cell will distribute this content to these multiple users who have interest in this content. That is to say: as multiple users in the community may have the same interest in the content, content centric delivery of one content can satisfy the interest of multiple users, and the total traffic of content delivery between the original server and the small cell can be reduced.

User ID versus Interest Based Content ID In MSNs, the content is produced by a user who has a social ID. The content ID can be a hierarchical name containing the mobile user ID. If multiple replicas of this content are delivered over different communities by using the content ID, these replicas can be recognized as the replicas of the same content. When a user wants to update content or publish some new information related to this content, the existing networks manage the replicas in

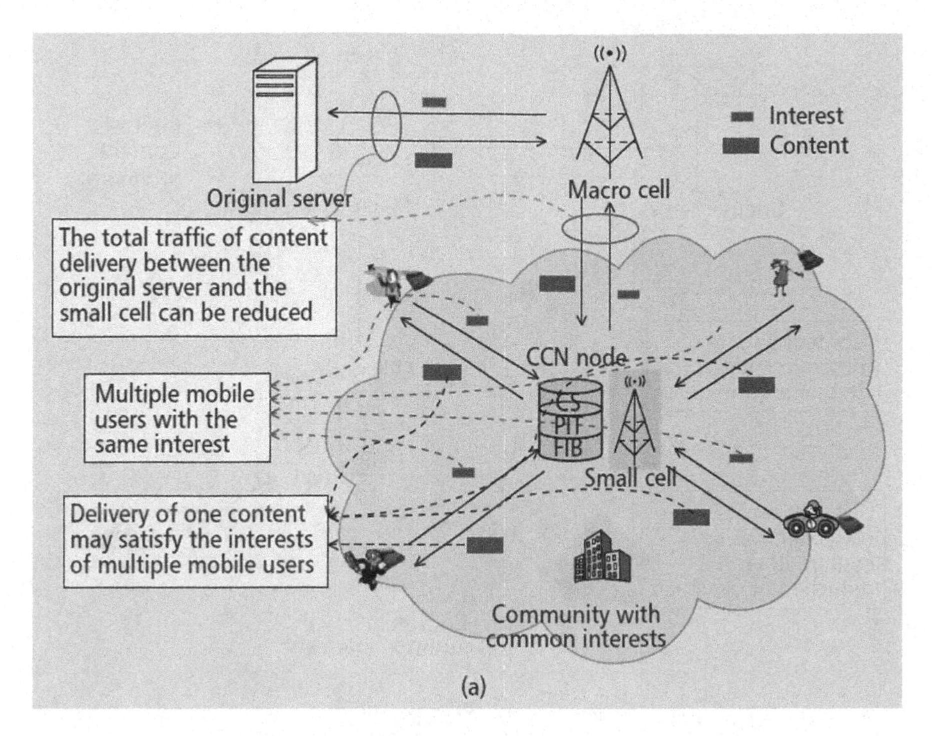

Fig. 6.3 Community with common interest versus interest based content

different IP addresses as different content and update these replicas with a high overhead. However, these replicas can be recognized as the replicas of the same content as shown in Fig. 6.4, with content ID these replicas can be recognized as the replicas of the same content. When the original content is updated, different replicas of this content in the same small cell can be updated on time by the command from the small cell, with a reduction of overhead.

Mobility versus Content Store A user may keep moving during the request of one content. If this user can only get the interested content from the original server, when he leaves the coverage area of one small cell, the connection between the original server and the small cell may be interrupted. Then, the content store in the CCNs nodes in small cells keeps some replicas of content. If the interested content is available in the content store, this replica can be provided to the user, as shown in Fig. 6.5. Therefore, if the user keeps moving through different small cells, he can get the interested content in the small cell without contacting to the original server, resulting in a reduction of delay.

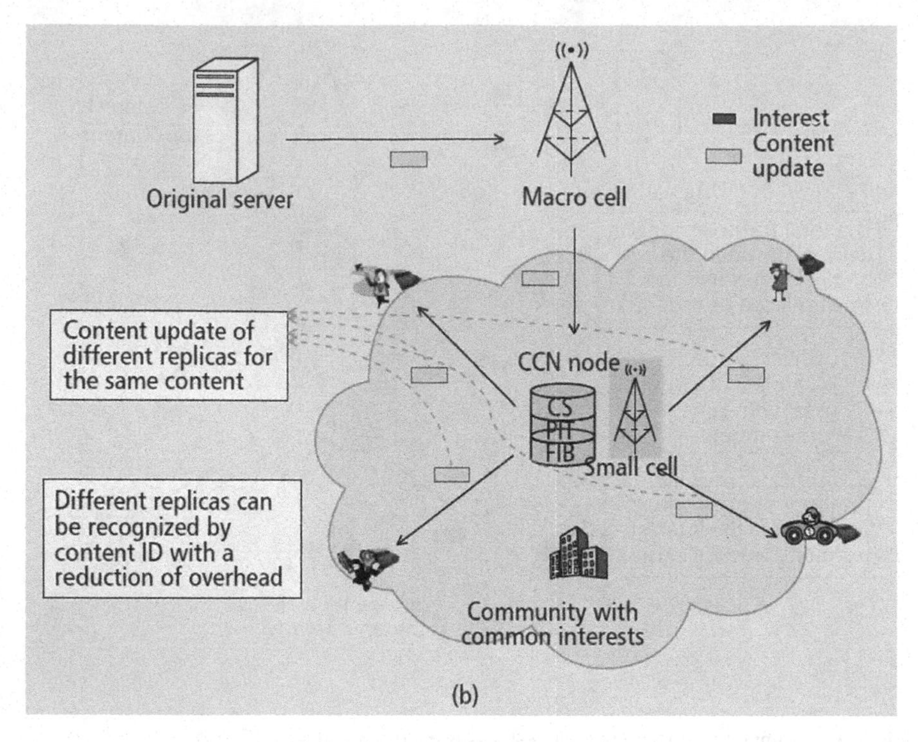

Fig. 6.4 User ID versus interest based content ID

6.4 Architecture of Content Centric Mobile Content Delivery

Based on the above introduction, we show a framework of content centric mobile content delivery in 5G in this section.

6.4.1 Network Architecture

Users with Community Users take mobile devices to join MSNs, where the community consists of different users with common interests. Users may deliver their own content to others who have social ties with them. They can also request content stored outside of the communities.

Small Cells Small cells are low power cell towers that are installed by the operator. Compared with macro cells, although the backhaul and access features are the same, the transmission power and coverage are lower. Generally, the small cell is directly responsible for communication among users who want to share content with each other based on their social ties in the community.

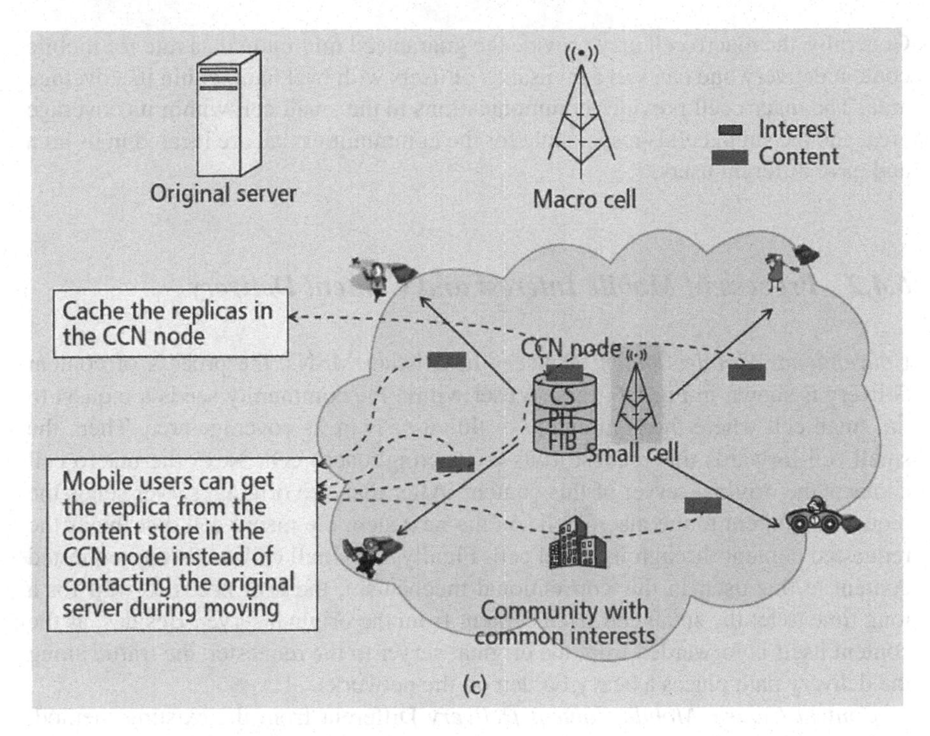

Fig. 6.5 Mobility versus content store

CCN Nodes CCN nodes are placed in the small cell for content centric delivery. Each CCN node has the following three parts with different data structures: content store, pending interest table, and forwarding information base.

- *Content Store (CS)* In each CCN node there is a caching space called the content store to keep replicas of some content. In fact, the content store has a limited capacity and cannot store all replicas of all content. The content store is similar to the buffer memory of conventional routers, but it is intended to cache content based on interest and has no direct relation with where this content is from.
- *Pending Interest Table (PIT)* The PIT is intended to control the pending interests that are currently waiting for responses. The track of forwarded interests to the destination can be kept, so the returned content can be sent to the content requester later. The PIT may keep a group of entries, and each entry will be erased as long as this entry is used to forward a matching content.
- *Forwarding Information Base (FIB)* The FIB is intended to forward interest to potential sources of matching content. Note that it can have multiple potential sources, so the query process can be parallel distributed. It is different from conventional routers, which only allow a single source.

Macro Cells The macro cell consists of the traditional operator-installed base stations. Because it covers a wide area, it may provide access on the order of kilometers.

Generally, the macro cell may provide the guaranteed minimum data rate for mobile content delivery and can serve thousands of users with backhaul within its coverage area. The macro cell provides communications to the small cell within its coverage area, and the small cell is responsible for the communities that are located in its area and have different users.

6.4.2 Process of Mobile Interest and Content Delivery

Conventional Mobile Content Delivery In existing MSNs, the process of content delivery is shown in Fig. 6.6. First, a user within the community sends a request to the small cell where the community of this user is in its coverage area. Then, the small cell forwards this request to its connecting macro cell. Next, the macro cell contacts the original server of this content. After that, the original server sends the requested content to the macro cell. As the next step, the macro cell distributes the requested content through its small cell. Finally, the small cell sends the requested content to this user. In this conventional mechanism, the user needs to wait for a long time to let the small cell fetch content from the original server. Besides, as the content itself is forwarded from the original server to the requester, the traffic along the delivery path places a heavy burden on the networks.

Content Centric Mobile Content Delivery Different from the existing method, content centric MSNs deliver content based on the interest. Users request content by broadcasting the interest in content. Any user who hears the interest and has the content can respond with the content. Therefore, during the process of content query, only the interest is forwarded.

According to the availability of replicas of the interested content in CCN node, there are two cases as follows.

Case 1: A Replica of the Content is Available in the Content Store of the CCN Node As shown in Fig. 6.7, when a user sends an interest of content, the interest will be sent to the content store of the CCN node in the small cell, where the community of this user is in its coverage. If the content store of the CCN node in the small cell keeps the replica of the content that the interest refers to, the replica of this content in the content store will be directly sent to the user.

Case 2: A Replica of the Content is Unavailable in the Content Store of the CCN Node In Fig. 6.8, if there is no replica of content in the content store that matches the interest, the CCN node will check whether there is a matching entry of the interest in the PIT. If there is such an exact-match entry in the PIT, it means that this is a pending interest. If such an entry is not available in the PIT either, the CCN node will check if this interest has an exact-match entry in the FIB. Then the CCN node will wait for the content to be fetched from a delivery path suggested by the FIB.

Based on the above two cases, if a user can cache its interested contents in the content store, both the response delay and the network traffic can be reduced. However, because of the limited capacity, not all of the replicas of content can be cached in the content store. If a newly arriving content needs to be cached in the content

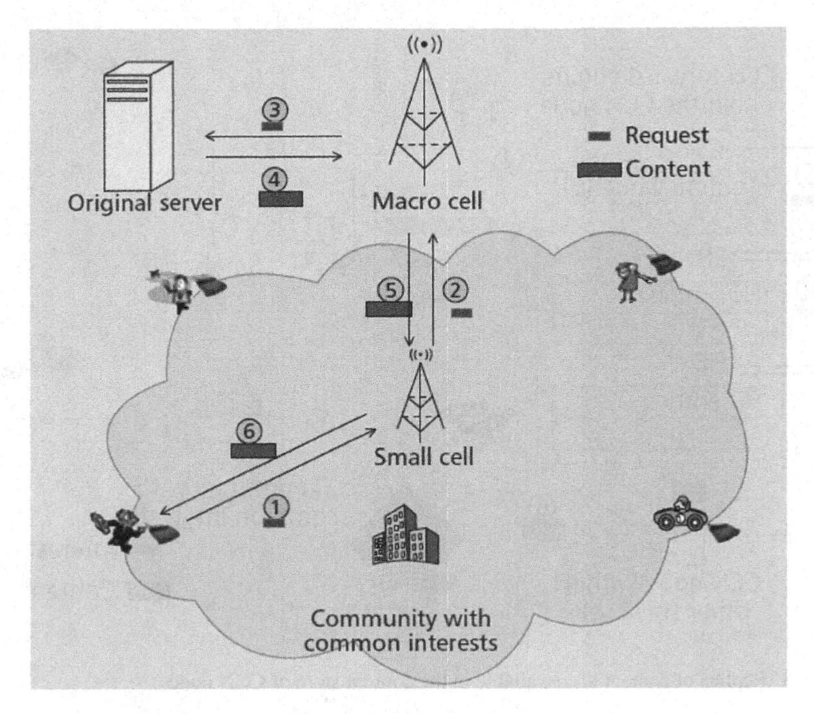

Fig. 6.6 Conventional mobile content delivery

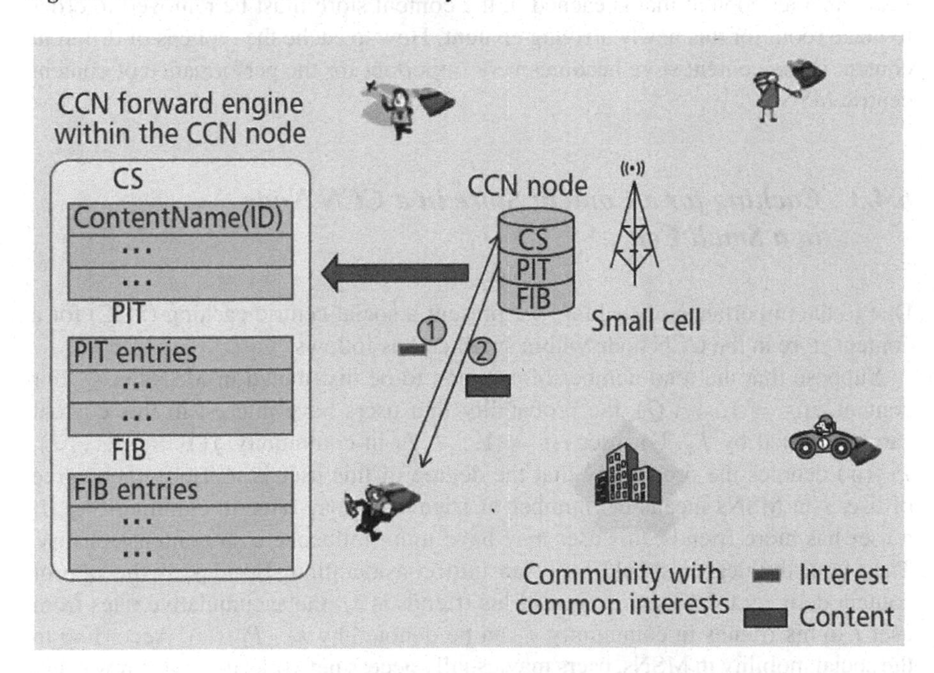

Fig. 6.7 Replica of content is available in the content store of CCN node

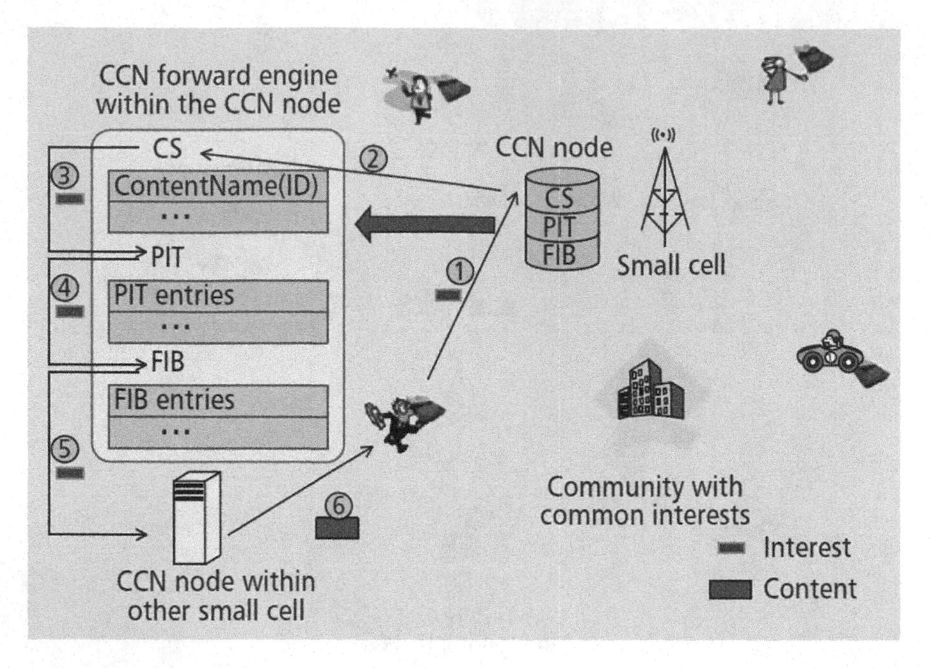

Fig. 6.8 Replica of content is unavailable in the content store of CCN node

store, another content that is cached in the content store must be removed in order to make room for this newly arriving content. How to cache the replicas of different content in the content store becomes very important for the performance of content centric MSNs.

6.4.3 Caching for a Content Store in a CCN Node in a Small Cell

Due to the importance of caching, we present a social centric caching (SCC) for a content store in the CCN node within small cell as follows.

Suppose that the total number of contents to be distributed in MSNs is Q. For content $q(q = 1, \ldots, Q)$, the probability that users have interest in this content can be defined by P_q. For user $i(i = 1, \ldots, I)$ in community $j(j = 1, \ldots, J)$, $P_{i,j}(w)$ denotes the probability that the degree of this user is w. Here, the degree of user i in MSNs means the number of friends he may have in community j. If a user has more friends, this user may have more influence over content delivery. Therefore, the degree should be taken into consideration. Besides, if the rate of content delivery from user i to one of his friends is λ_i, the accumulative rates from user i to his friends in community j can be denoted by $\lambda_i \cdot P_{i,j}(w)$. According to the social mobility in MSNs, users may usually access networks through a few small cells. For example, students at a university may always access networks from the small cells in the dormitory and classroom. For user i, if he has connected to a small cell $k(k = 1, \ldots, K)$ by M times during the access of MSNs, as for the m-th time $(m = 1, \ldots, M)$, the connection time is denoted by $S_{i,m,k}$. We call $S_{i,m,k}$ the social stay of user i during the m-th connection with small cell k. Then the probability that user i stays in the coverage area of small cell k becomes $T_{stay}^{i,k} = \frac{\sum_m S_{i,m,k}}{T}$, where T is a watching period.

If an interest in content q arises from user i, when the replica of content q is being cached in the content store of a CCN node within the small cell k, the cost in delay to fetch the content q from this content store to user i can be defined by $d_{i,k,q}$. Otherwise, when the replica of content q is unavailable in the content store of a CCN node within the small cell k, the CCN node needs to check both the PIT and the FIB to fetch the replica from another CCN node that is not located in small cell k, resulting in a cost denoted by $d^{'}_{i,k,q}$.

Then, we can obtain the relative cost if content q is not being cached in the content store of a CCN node within small cell k by $cost_{q,k} = \sum_i \sum_j (P_q \cdot \lambda_i \cdot P_{i,j}(w) \cdot T_{stay}^{i,k} \cdot (d^{'}_{i,k,q} - d_{i,k,q}))$.

6.4.4 Performance Evaluation

We compare the hit ratio of the proposed SCC with the random method in Fig. 6.9. In the simulation, the distribution of interests follows the Zip-f distribution [24]. The rate of content delivery is determined by the Poison distribution. The degree

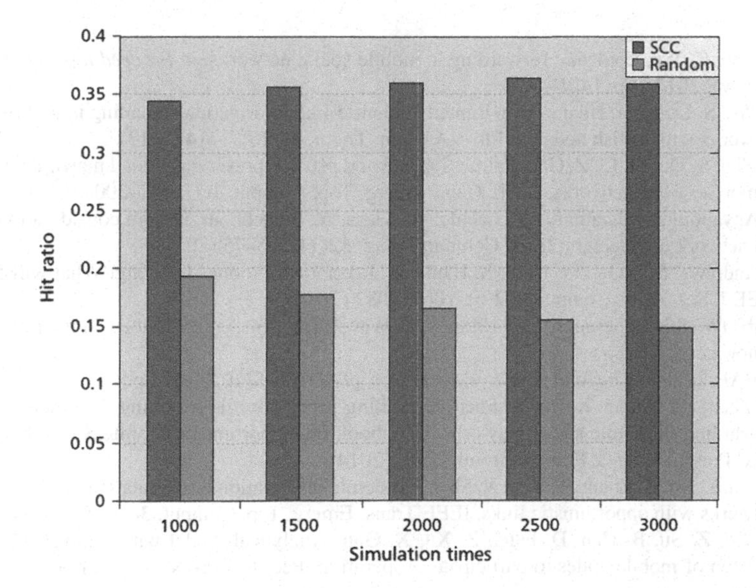

Fig. 6.9 Hit ratio with different simulation times

of users is decided by the power law distribution. Here, the random method means that the priorities of replicas to cache in the content store are determined at random. We carried out experiments to test the performance under different simulation times. Here, the priorities of replicas to cache in the content store are determined by $cost_{q,k}$. As $cost_{q,k}$ takes into consideration the popularity of content, features of the social community, and delay, it can be looked upon as the cost to remove cached content from the content store. From Fig. 6.9, it can be obtained that the proposed scheme can result in a better hit ratio than the random method. More details can be seen in [25].

6.5 Summary

In this chapter, a novel framework to deliver content over content centric MSNs in 5G has been presented. We have shown the system structure consisting of users, CCN nodes, small cells, and macro cells. The process of mobile content delivery based on the interest of content and users in communities has been discussed. As caching in the content store of a CCN node plays an important role in the performance of content delivery, we also propose a caching proposed scheme to determine which replicas should be stored in the content store. Simulation results have shown the efficiency of our proposed model.

References

1. H. Sun, C. Wu, Epidemic forwarding in mobile social networks, in *Proceedings of IEEE ICC* (Ottawa, 2012), pp. 1421–1425
2. Y. Wu, S. Deng, H. Huang, Hop limited epidemic-like information spreading in mobile social networks with selfish nodes. J. Phys. A Math. Theor. **46**(26), 1–14 (2013)
3. H. Zhu, S. Du, M. Li, Z. Gao, Fairness-aware and privacy-preserving friend matching protocol in mobile social networks. IEEE Trans. Emerg. Top. Comput. **1**(1), 192–200 (2013)
4. P. Agyapong, M. Iwamura, D. Staehle, W. Kiess, A. Benjebbour, Design considerations for a 5G network architecture. IEEE Commun. Mag. **52**(11), 65–75 (2014)
5. J. Andrews, S. Buzzi, W. Choi, S. Hanly, A. Lozano, A. Soong, J. Zhang, What will 5G be? IEEE J. Sel. Areas Commun. **32**(6), 1065–1082 (2014)
6. 5GNOW: 5th Generation Non-orthogonal Waveforms for Asynchronous Signalling. www.5gnow.eu
7. 20BAH:2020 and beyond adhoc. www.arib.or.jp/ADWICS/2020bah-J.pdf
8. K. Zhang, X. Liang, R. Lu, X. Shen, Exploiting private profile matching for efficient packet forwarding in mobile social networks. Handbook on Opportunistic Mobile Social Networks, CRC Press, Taylor & Francis Group, USA, (2014)
9. Q. Xu, Z. Su, K. Zhang, P. Ren, X. Shen, Epidemic information dissemination in mobile social networks with opportunistic links. IEEE Trans. Emerg. Top. Comput. **3**(3), 399–409 (2015)
10. Q. Xu, Z. Su, B. Han, D. Fang, Z. Xu, X. Gan, Analytical model with a novel selfishness division of mobile nodes to participate cooperation. Peer-to-Peer Netw. Appl. **9**(4), 712–720 (2016)

11. P. Costa, C. Mascolo, M. Musolesi, G. Picco, Socially-aware routing for publish-subscribe in delay-tolerant mobile ad hoc networks. IEEE J. Sel. Areas Commun. **26**(5), 748–760 (2008)
12. W. Zhang, Y. Ye, H. Tan, Q. Dai, T. Li, Information diffusion model based on social network, in *Proceedings of ICMCS AAISC* (Berlin, 2010), pp. 145–450
13. K. Zheng, F. Hu, W. Wang, W. Xiang, M. Dohler, Radio resource allocation in LTE-advanced cellular networks with M2M communications. IEEE Commun. Mag. **50**(7), 184–192 (2012)
14. V. Jacobson, D. Smetters, J. Thornton, P. Plass, N. Briggs, R. Braynard, Networking named content, in *Proceedings of the ENET* (Rome, 2009), pp. 1–12
15. M. Ohtani, K. Tsukamoto, Y. Koizumi, H. Ohsaki, M. Imase, K. Hato, J. Murayama, VCCN: virtual content-centric networking for realizing group-based communication, in *Proceedings of the IEEE ICC* (Budapest, 2013), pp. 3476–3480
16. A. Detti, S. Salsano, N. Blefari-Melazzi, IP protocol suite extensions to support CONET information centric networking, in *IETF, draft-detti-conet-ip-option*. http://tools.ietf.org/html/draft-detti-conet (2013)
17. Z. Su, Q. Xu, H. Zhu, Y. Wang, A novel design for content delivery over software defined mobile social networks. IEEE Netw. **29**(4), 62–67 (2015)
18. L. Lei, Y. Zhang, X. Shen, C. Lin, Z. Zhong, Performance analysis of device-to-device communications with dynamic interference using stochastic Petri nets. IEEE Trans. Wirel. Commun. **12**(12), 6121–6141 (2013)
19. X. Sheng, J. Tang, X. Xiao, G. Xue, Leveraging GPS-less sensing scheduling for green mobile crowd sensing. IEEE Internet Things J. **1**(4), 328–336 (2014)
20. E. Bulut, B. Szymanski, Friendship based routing in delay tolerant mobile social networks. in *Proceedings of the IEEE GLOBECOM* (Miami, 2010), pp. 1–5
21. M. Li, S. Yu, N. Cao, W. Lou, Privacy-preserving distributed profile matching in proximity-based mobile social networks. IEEE Trans. Wirel. Commun. **12**(5), 2024–2033 (2013)
22. X. Jiang, J. Bi, Interest set mechanism to improve the transport of named data networking, in *Proceedings of the ACM SIGCOMM* vol. 43, no. 4 (New York, 2013), pp. 515–516
23. S. Wang, J. Bi, J. Wu, On performance of cache policy in information-centric networking, in *Proceedings of the ICCCN* (Munich, 2012), pp. 1–7
24. L. Breslau, P. Cao, L. Fan, G. Phillips, S. Shenker, Web caching and Zipf-like distributions: evidence and implications, in *Proceedings of the IEEE INFOCOM*, vol. 1 (New York, 1999), pp. 126–134
25. Z. Su, Q. Xu, Content distribution over content centric mobile social networks in 5G. IEEE Commun. Mag. **53**(6), 66–72 (2015)

Chapter 7
Conclusions and Future Directions

In this chapter, we summarize the monograph and provide future research directions.

7.1 Conclusions

In this monograph, we have investigated the modeling and optimization for MSNs in the following aspects.

- *Modeling and Optimization of Epidemic Information Dissemination in MSNs*
  In MSNs, users adopt a store-carry-and-forward mode to disseminate information with the opportunistic links. Due to the dynamical properties of opportunistic links, the procedures of information dissemination becomes hard to predict. An analytical model is developed to optimize the epidemic information dissemination in MSNs. The change of user's interests is studied with two novel elements: pre-immunity and immunity. By introducing four rules for epidemic information dissemination, the analytical model is developed based on ordinary differential equations. The extensive trace-driven simulations demonstrated the efficiency of the analytical model.
- *Modeling of Selfishness-Aware Incentive Mechanism in MSNs*
  Due to the limited resources (e.g., battery and buffer), not all users are willing to contribute information to MSNs. This selfish behavior may affect the performance of mobile social services, such as mobile crowd sensing. In order to encourage the individuals to participate in the MSN cooperation, an incentive model is developed considering both weak selfishness and extreme selfishness. The impact of selfishness has been analyzed by using the proposed model. The simulation based on the real trace shows the effects of weak selfishness and extreme selfishness. The model is optimal and stimulates users to contribute information to MSNs.
- *Optimization of Relay Service in MSNs*
  In MSNs, users can share their content with each other by using mobile devices equipped with short-range wireless interfaces via peer-to-peer opportunistic links.

Z. Su et al., *Modeling and Optimization for Mobile Social Networks*,
DOI 10.1007/978-3-319-47922-4_7

To efficiently deliver content from the source to the destination, user-assisted relay is essential for the content delivery in a store-carry-forward way. We introduce a bargain game to model the transaction pricing between the bundle carrier and the relay node. An optimal relay selection scheme is proposed to encourage users to participate in bundle delivery.

- *Modeling and Optimization of Cloud Resource Allocation in MSNs*
  Media cloud has been advocated to deploy cloud resource to provide users with multimedia services by allocating some resource through the brokers. Due to the different resource requirements from users, the optimal allocation of the limited resource becomes important. Based on the competition among media cloud, brokers and users, an optimal resource allocation scheme is proposed. The media cloud determines the price to lease cloud resource to the nearby brokers. After obtaining the cloud resource based on the price provided by the media cloud, the broker allocates the resource to users. We formulate the resource allocation problem as a Stackelberg game, and propose an optimal resource allocation model where users can selectively connect brokers to obtain resources. Simulation results show the efficiency of the proposed optimal resource allocation model.
- *Modeling and Optimization of Heterogeneous MSN Architecture*
  The existing network architecture is based on the design that relies on the IP address of the content. With the rapid spread of social network applications, users may use various types of mobile devices to access MSNs by different wireless connections. The optimal model of a heterogeneous network architecture is needed for the current MSNs. By introducing a new framework consisting of mobile users, content-centric nodes, small cells, and macro cells, an optimal network architecture is proposed. The content can be managed in a content-centric mode, where the content is recognized by naming information instead of the conventional IP. Simulation results show that the content can be efficiently delivered with the proposed optimal network architecture.

7.2 Future Research Directions

Although this monograph presents some important research topics of modeling and optimization for MSNs, there are still several open problems including but not limited to the followings.

7.2.1 *Big Data and MSNs*

In MSNs, massive data are generated by users for social activities by using mobile devices. The average amount of mobile data traffic per month in 2015 was 4.2 EB/mo [1]. Among the above traffic, there is a very large proportion caused by video and audio. In 2019, mobile data traffic is predicted to be an almost sixfold increase which will be 24.3 EB/mo. Mobile social data have the properties including the large volume, wide variety, fast velocity, and economic value which are called “4Vs” [2].

In addition to the existing “4Vs”, mobile big data in MSNs has some unique and important features including aggregate features and individual features. These features can be captured by some analytical methods such as data mining and machine learning. For aggregate features, it can be exploited to improve the efficiency of MSNs. For example, the wireless resource can be allocated based on the demand of users in the communities that have the same interest. For individual features, it is useful for network operators to provide the improved and personalized services. For example, social relations and mobile patterns of mobile users can be used to design an optimal content routing.

Mobile big data applications are helpful to provide users with better social services. For example, based on the preferences of users in a community, the content recommendation system can be built. The content which users may have interest can be delivered based on a publish/subscribe mode. How to apply the big data applications into MSNs is still a challenge.

7.2.2 *IoTs and MSNs*

The number of mobile devices has been larger than the world population since 2014. Internet of things (IoTs) [3] make users, data and devices being linked with each other. The integration of MSNs with the IoTs results in the mobile social internet of things (MSIoTs), as more and more users can take mobile devices to exchange data in IoTs. Based on the social tie of their owners, the devices in MSIoTs can establish social relationships in an autonomous way.

The MSIoTs connect objects of the physical world with users. Without a central controller. The MSIoTs can bring new possibilities for data delivery in a predominantly self-organized way. For example, the driving car in the street can search information and acquire data not only from Internet, but also from other driving cars, passengers, and nearby mobile devices.

Instead of working as an independent unit, multiple devices may work cooperatively based on their social relationships to execute a task. For example, when a group of vehicles with sensors work together to collect data in an urban area, the social relationship among vehicles should be studied according to their drivers’, in order to determine the optimal route, motorcade and velocity. The integration of MSNs with IoTs has a big potential in both academic and industry.

7.2.3 *Content Centric MSNs*

The current network architecture in MSNs is location-based, where the content is delivered to users based on their IP. Due to the ever expanding scale of MSNs, there are multiple replicas of the same content which is stored at different locations. For example, two users may keep the replica of the same content downloaded from different sites. As these two replicas own different IPs, they may be recognized as different content, resulting in a mistaken content exchange. In fact, users care more about what this content is than where this content is from. Content centric networks have been advocated to manage contents based on their naming information [4, 5]. The content is requested by sending interests in it from users.

In content centric MSNs, the content is recognized by a hierarchical name containing its content ID. When multiple replicas of the same content are delivered to different users in different communities, these replicas can be identified from the same content. The mistakes of content exchange and the efficiency to manage different replicas can be expected to reduce.

In addition, the content that is frequently requested by users in certain community can be stored at the nearby content cache node, which can reduce the delay for users to obtain the content in this community. Thus, how to design a content-centric MSNs becomes an open problem.

7.2.4 Emerging Applications in MSNs

Many MSN applications are expected to emerge in the future. Recently, mobile-assisted learning [6] has been realized by using the blackboard mobile tools consisting of Apple and Android devices. By incorporating mobile networks in the teaching system, mobile-assisted learning can provide students with a novel social-based learning platform, and encourage collaboration and discussions. To make the efficient education, mobile-assisted learning needs the optimal analysis based on the properties of learners in MSNs. For example, the optimal community construction to make learning group by considering the student's score, interest, motivation, etc.

MSNs can also be considered as user-centric networks to support various services for users. For example, the MSNs can be applied to healthcare support. The physical condition of patients can be monitored by smart mobile, which can deliver the health report to family or hospital. There are also some other applications such as providing game services for users. After downloading the game applications, a group of users can play game in a distributed mode. They can form the MSNs through device-to-device (D2D) connection without a base station. It can be predicted that more and more applications with MSNs will emerge in the future.

In conclusion, we hope this monograph sheds more light on the importance of modeling and optimization for MSNs, which still requires further research effort along the emerging line.

References

1. Cisco visual networking index: global mobile data traffic forecast update 2014–2019, in *Cisco Public Information* (2014). http://www.cisco.com/c/en/us/solutions/service-provider/visual-networking-index-vni/index.html
2. H. Hu, Y. Wen, T. Chua, X. Li, Toward scalable systems for big data analytics: a technology tutorial. IEEE Access **2**, 652–687 (2014)
3. B. Cheng, D. Zhu, S. Zhao, J. Chen, Situation-aware IoT service coordination using the event-driven SOA paradigm. IEEE Trans. Netw. Serv. Manag. **13**(2), 349–361 (2016)
4. Z. Su, Q. Xu, Content distribution over content centric mobile social networks in 5G. IEEE Commun. Mag. **53**(6), 66–72 (2015)
5. A. Detti, S. Salsano, N. Blefari-Melazzi, IP protocol suite extensions to support conet information centric networking, in *IETF, draft-detti-conet-ip-option*. http://tools.ietf.org/html/draft-detti-conet (2013)
6. I. Simonova, P. Poulova, Social networks and mobile devices in higher education: pilot project, in *Proceedings of the COMPSAC* (Taichung, 2013), pp. 851–856

References

[illegible]

Printed in the United States
By Bookmasters